Fatigue

The McGraw-Hill *CONTROLLING PILOT ERROR* Series

Weather
Terry T. Lankford

Communications
Paul E. Illman

Automation
Vladimir Risukhin

Controlled Flight into Terrain (CFIT/CFTT)
Daryl R. Smith

Training and Instruction
David A. Frazier

Checklists and Compliance
Thomas P. Turner

Maintenance and Mechanics
Larry Reithmaier

Situational Awareness
Paul A. Craig

Fatigue
James C. Miller

Culture, Environment, and CRM
Tony Kern

Cover Photo Credits (clockwise from upper left): PhotoDisc; Corbis Images; from *Spin Management and Recovery* by Michael C. Love; PhotoDisc; PhotoDisc; PhotoDisc; image by Kelly Parr; © 2001 Mike Fizer, all rights reserved; *Piloting Basics Handbook* by Bjork, courtesy of McGraw-Hill; PhotoDisc.

CONTROLLING PILOT ERROR

Fatigue

James C. Miller

McGraw-Hill
New York   Chicago   San Francisco   Lisbon   London   Madrid
Mexico City   Milan   New Delhi   San Juan   Seoul
Singapore   Sydney   Toronto

Cataloging-in-Publication Data is on file with the Library of Congress

McGraw-Hill
A Division of The McGraw·Hill Companies

Copyright © 2001 by The McGraw-Hill Companies, Inc. All rights reserved. Printed in the United States of America. Except as permitted under the United States Copyright Act of 1976, no part of this publication may be reproduced or distributed in any form or by any means, or stored in a data base or retrieval system, without the prior written permission of the publisher.

1 2 3 4 5 6 7 8 9 0   DOC/DOC   0 7 6 5 4 3 2 1

ISBN 0-07-137412-4

The sponsoring editor for this book was Shelley Ingram Carr, the editing supervisor was Steven Melvin, and the production supervisor was Sherri Souffrance. It was set in Garamond per the TAB3A design by Victoria Khavkina of McGraw-Hill's Professional Book Group composition unit.

Printed and bound by R. R. Donnelley & Sons Company.

This book is printed on recycled, acid-free paper containing a minimum of 50% recycled de-inked fiber.

McGraw-Hill books are available at special quantity discounts to use as premiums and sales promotions, or for use in corporate training programs. For more information, please write to the Director of Special Sales, Professional Publishing, McGraw-Hill, Two Penn Plaza, New York, NY 10121-2298. Or contact your local bookstore.

Information contained in this work has been obtained by The McGraw-Hill Companies, Inc. ("McGraw-Hill") from sources believed to be reliable. However, neither McGraw-Hill nor its authors guarantee the accuracy or completeness of any information published herein and neither McGraw-Hill nor its authors shall be responsible for any errors, omissions, or damages arising out of use of this information. This work is published with the understanding that McGraw-Hill and its authors are supplying information but are not attempting to render engineering or other professional services. If such services are required, the assistance of an appropriate professional should be sought.

Contents

Series Introduction *vii*
Foreword *xvii*
1 Introduction *1*
2 It's Been a Rough Day *17*
3 The Daily Two-Peak Pattern of Errors *41*
4 Sleep Biology and Napping *55*
5 Sleep in the News *73*
6 Dealing with Jet Lag *93*
7 Cumulative Fatigue: It All Adds Up *109*
8 Can the Pharmacy Help? *125*
9 A Prescription for Fighting Fatigue *141*
References *153*
Appendix A: Hints for Good Sleep Hygiene *159*
Appendix B: Napping Plan *165*
Index *169*

Series Introduction

The Human Condition

The Roman philosopher Cicero may have been the first to record the much-quoted phrase "to err is human." Since that time, for nearly 2000 years, the malady of human error has played out in triumph and tragedy. It has been the subject of countless doctoral dissertations, books, and, more recently, television documentaries such as "History's Greatest Military Blunders." Aviation is not exempt from this scrutiny, as evidenced by the excellent Learning Channel documentary "Blame the Pilot" or the NOVA special "Why Planes Crash," featuring John Nance. Indeed, error is so prevalent throughout history that our flaws have become associated with our very being, hence the phrase *the human condition*.

The Purpose of This Series

Simply stated, the purpose of the Controlling Pilot Error series is to address the so-called human condition, improve performance in aviation, and, in so doing, save a few lives. It is not our intent to rehash the work of over a millennia of expert and amateur opinions but rather to *apply* some of the more important and insightful theoretical perspectives to the life and death arena of manned flight. To the best of my knowledge, no effort of this magnitude has ever been attempted in aviation, or anywhere else for that matter. What follows is an extraordinary combination of why, what, and how to avoid and control error in aviation.

Because most pilots are practical people at heart— many of whom like to spin a yarn over a cold lager—we will apply this wisdom to the daily flight environment, using a case study approach. The vast majority of the case studies you will read are taken directly from aviators who have made mistakes (or have been victimized by the mistakes of others) and survived to tell about it. Further to their credit, they have reported these events via the anonymous Aviation Safety Reporting System (ASRS), an outstanding program that provides a wealth of extremely useful and *usable* data to those who seek to make the skies a safer place.

A Brief Word about the ASRS

The ASRS was established in 1975 under a Memorandum of Agreement between the Federal Aviation Administration (FAA) and the National Aeronautics and Space Administration (NASA). According to the official ASRS web site, *http://asrs.arc.nasa.gov*

- The ASRS collects, analyzes, and responds to voluntarily submitted aviation safety incident reports in order to lessen the likelihood of aviation accidents. ASRS data are used to:
  - Identify deficiencies and discrepancies in the National Aviation System (NAS) so that these can be remedied by appropriate authorities.

- Support policy formulation and planning for, and improvements to, the NAS.
- Strengthen the foundation of aviation human factors safety research. This is particularly important since it is generally conceded *that over two-thirds of all aviation accidents and incidents have their roots in human performance errors* (emphasis added).

Certain types of analyses have already been done to the ASRS data to produce "data sets," or prepackaged groups of reports that have been screened "for the relevance to the topic description" (ASRS web site). These data sets serve as the foundation of our Controlling Pilot Error project. The data come *from* practitioners and are *for* practitioners.

The Great Debate

The title for this series was selected after much discussion and considerable debate. This is because many aviation professionals disagree about what should be done about the problem of pilot error. The debate is basically three sided. On one side are those who say we should seek any and all available means to *eliminate* human error from the cockpit. This effort takes on two forms. The first approach, backed by considerable capitalistic enthusiasm, is to automate human error out of the system. Literally billions of dollars are spent on so-called human-aiding technologies, high-tech systems such as the Ground Proximity Warning System (GPWS) and the Traffic Alert and Collision Avoidance System (TCAS). Although these systems have undoubtedly made the skies safer, some argue that they have made the pilot more complacent and dependent on the automation, creating an entirely new set of pilot errors. Already the automation enthusiasts are seeking robotic answers for this new challenge. Not surprisingly, many pilot trainers see the problem from a slightly different angle.

Another branch on the "eliminate error" side of the debate argues for higher training and education standards, more accountability, and better screening. This group (of which I count myself a member) argues that some industries (but not yet ours) simply don't make serious errors, or at least the errors are so infrequent that they are statistically nonexistent. This group asks, "How many errors should we allow those who handle nuclear weapons or highly dangerous viruses like Ebola or anthrax?" The group cites research on high-reliability organizations (HROs) and believes that aviation needs to be molded into the HRO mentality. (For more on high-reliability organizations, see "Culture, Environment, and CRM" in this series.) As you might expect, many status quo aviators don't warm quickly to these ideas for more education, training, and accountability—and point to their excellent safety records to say such efforts are not needed. They recommend a different approach, one where no one is really at fault.

On the far opposite side of the debate lie those who argue for "blameless cultures" and "error-tolerant systems." This group agrees with Cicero that "to err is human" and advocates "error-management," a concept that prepares pilots to recognize and "trap" error before it can build upon itself into a mishap chain of events. The group feels that training should be focused on primarily error mitigation rather than (or, in some cases, in addition to) error prevention.

Falling somewhere between these two extremes are two less-radical but still opposing ideas. The first approach is designed to prevent a recurring error. It goes something like this: "Pilot X did this or that and it led to a mishap, so don't do what Pilot X did." Regulators are particularly fond of this approach, and they attempt to regulate the last mishap out of future existence. These so-called rules written in blood provide the traditionalist with plenty of training materials and even come with ready-made case studies—the mishap that precipitated the rule.

Opponents to this "last mishap" philosophy argue for a more positive approach, one where we educate and train *toward* a complete set of known and valid competencies (positive behaviors) instead of seeking to eliminate negative behaviors. This group argues that the professional airmanship potential of the vast majority of our aviators is seldom approached—let alone realized. This was the subject of an earlier McGraw-Hill release, *Redefining Airmanship*.[1]

Who's Right? Who's Wrong? Who Cares?

It's not about *who's* right, but rather *what's* right. Taking the philosophy that there is value in all sides of a debate, the Controlling Pilot Error series is the first truly comprehensive approach to pilot error. By taking a unique "before-during-after" approach and using modern-era case studies, 10 authors—each an expert in the subject at hand—methodically attack the problem of pilot error from several angles. First, they focus on error prevention by taking a case study and showing how preemptive education and training, applied to planning and execution, could have avoided the error entirely. Second, the authors apply error management principles to the case study to show how a mistake could have been (or was) mitigated after it was made. Finally, the case study participants are treated to a thorough "debrief," where alternatives are discussed to prevent a reoccurrence of the error. By analyzing the conditions before, during, and after each case study, we hope to combine the best of all areas of the error-prevention debate.

A Word on Authors and Format

Topics and authors for this series were carefully analyzed and hand-picked. As mentioned earlier, the topics were taken from preculled data sets and selected for their relevance by NASA-Ames scientists. The authors were chosen for their interest and expertise in the given topic area. Some are experienced authors and researchers, but, more important, *all* are highly experienced in the aviation field about which they are writing. In a word, they are practitioners and have "been there and done that" as it relates to their particular topic.

In many cases, the authors have chosen to expand on the ASRS reports with case studies from a variety of sources, including their own experience. Although Controlling Pilot Error is designed as a comprehensive series, the reader should not expect complete uniformity of format or analytical approach. Each author has brought his own unique style and strengths to bear on the problem at hand. For this reason, each volume in the series can be used as a stand-alone reference or as a part of a complete library of common pilot error materials.

Although there are nearly as many ways to view pilot error as there are to make them, all authors were familiarized with what I personally believe should be the industry standard for the analysis of human error in aviation. The Human Factors Analysis and Classification System (HFACS) builds upon the groundbreaking and seminal work of James Reason to identify and organize human error into distinct and extremely useful subcategories. Scott Shappell and Doug Wiegmann completed the picture of error and error resistance by identifying common fail points in organizations and individuals. The following overview of this outstanding guide[2] to understanding pilot error is adapted from a United States Navy mishap investigation presentation.

> Simply writing off aviation mishaps to "aircrew error" is a simplistic, if not naive, approach to mishap causation. After all, it is well established that mishaps cannot be attributed to a single cause, or in most instances, even a single individual. Rather, accidents are the end result of a myriad of latent and active failures, only the last of which are the unsafe acts of the aircrew.
>
> As described by Reason,[3] active failures are the actions or inactions of operators that are believed to cause the accident. Traditionally referred to as "pilot error," they are the last "unsafe acts" committed by aircrew, often with immediate and tragic consequences. For example, forgetting to lower the landing gear before touch down or hotdogging through a box canyon will yield relatively immediate, and potentially grave, consequences.
>
> In contrast, latent failures are errors committed by individuals within the supervisory chain of command that effect the tragic sequence of events characteristic of an accident. For example, it is not difficult to understand how tasking aviators at the expense of quality crew rest can lead to fatigue and ultimately errors (active failures) in the cockpit. Viewed from this perspective then, the unsafe acts of aircrew are the end result of a long chain of causes whose roots originate in other parts (often the upper

echelons) of the organization. The problem is that these latent failures may lie dormant or undetected for hours, days, weeks, or longer until one day they bite the unsuspecting aircrew....

What makes the [Reason's] "Swiss Cheese" model particularly useful in any investigation of pilot error is that it forces investigators to address latent failures within the causal sequence of events as well. For instance, latent failures such

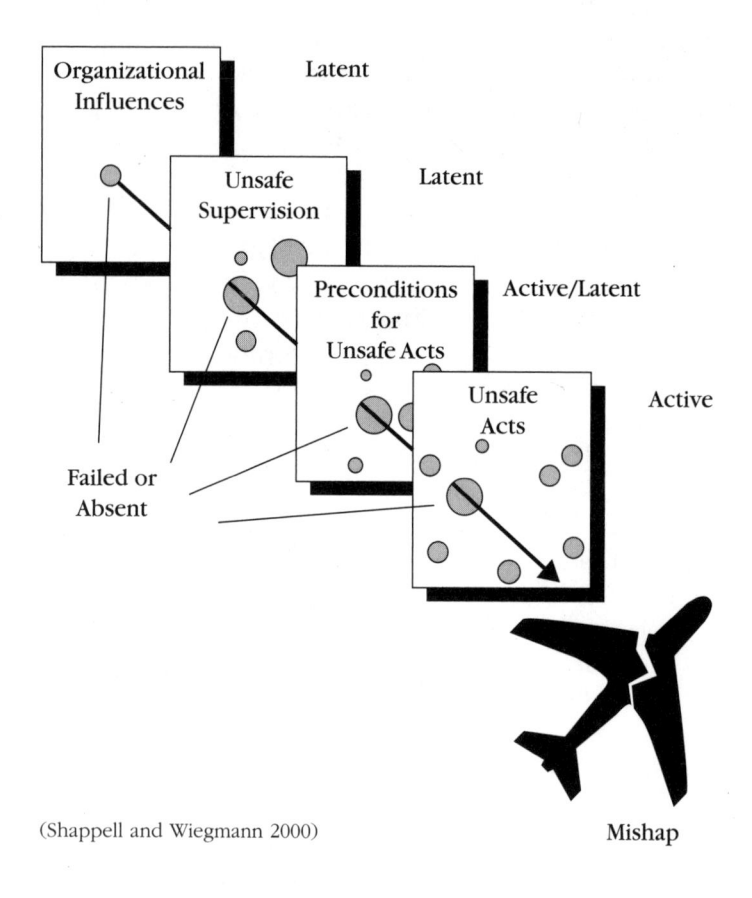

(Shappell and Wiegmann 2000)

as fatigue, complacency, illness, and the loss of situational awareness all effect performance but can be overlooked by investigators with even the best of intentions. These particular latent failures are described within the context of the "Swiss Cheese" model as preconditions for unsafe acts. Likewise, unsafe supervisory practices can promote unsafe conditions within operators and ultimately unsafe acts will occur. Regardless, whenever a mishap does occur, the crew naturally bears a great deal of the responsibility and must be held accountable. However, in many instances, the latent failures at the supervisory level were equally, if not more, responsible for the mishap. In a sense, the crew was set up for failure....

But the "Swiss Cheese" model doesn't stop at the supervisory levels either, the organization itself can impact performance at all levels. For instance, in times of fiscal austerity funding is often cut, and as a result, training and flight time is curtailed. Supervisors are therefore left with tasking "non-proficient" aviators with sometimes-complex missions. Not surprisingly, causal factors such as task saturation and the loss of situational awareness will begin to appear and consequently performance in the cockpit will suffer. As such, causal factors at all levels must be addressed if any mishap investigation and prevention system is going to work.[4]

The HFACS serves as a reference for error interpretation throughout this series, and we gratefully acknowledge the works of Drs. Reason, Shappell, and Wiegmann in this effort.

No Time to Lose

So let us begin a journey together toward greater knowledge, improved awareness, and safer skies. Pick up any volume in this series and begin the process of self-analysis that is required for significant personal or organizational change. The complexity of the aviation environment demands a foundation of solid airmanship and a healthy, positive approach to combating pilot error. We believe this series will help you on this quest.

References

1. Kern, Tony, *Redefining Airmanship,* McGraw-Hill, New York, 1997.

2. Shappell, S. A., and Wiegmann, D. A., *The Human Factors Analysis and Classification System—HFACS,* DOT/FAA/AM-00/7, February 2000.

3. Reason, J. T., *Human Error,* Cambridge University Press, Cambridge, England, 1990.

4. U.S. Navy, *A Human Error Approach to Accident Investigation,* OPNAV 3750.6R, Appendix O, 2000.

Tony Kern

Foreword

Mention fatigue to any pilot and you are sure to get a couple of good "there I was" war stories. In fact, fatigue has become a common denominator in most of Western society, a badge of honor for the workaholic. The problems associated with carrying this badge airborne are obvious.

A few years ago, I was passing through Cannon Air Force Base when I picked up the newspaper and read an article about a general aviation pilot who flew his last sortie. As I was reading this manuscript as the series editor, I went back and dug through my rat's nest of old articles and newspaper clippings and pulled it out. I have paraphrased the article below.

Darkness, pilot fatigue cited in June crash near Clovis

>Darkness and pilot fatigue played roles in the crash of a single-engine airplane in a cornfield near Clovis Municipal Airport, federal investigators said.
>
>The pilot, a 59-year-old man from Albuquerque died when the 1958 Cessna he was piloting clipped some power lines and crashed at about 12:30 A.M. June 28, 1998. He was a truck driver who drove for 12 hours in New Mexico and

Arizona the afternoon and evening before he took off from Los Lunas for Clovis, the National Transportation Safety Board reported Thursday.

The airplane hit the lines as it approached the airport, came in low over the runway and started to climb again, the report said. The aircraft then crashed.

The moon had set, and the pilot was not licensed to fly by the airplane's instruments alone, the NTSB said. The probable causes of the crash were the pilot's failure to clear the lines and his loss of control of the airplane as he tried to circle the airport for another landing attempt, the report said.

It struck me as odd that a non-instrument-rated pilot would be flying in near total darkness at midnight after a full day on the road. I now understand better how fatigue affects a pilot's judgment, and the purpose of this outstanding book is to assist you in doing the same.

Michael Mann, an administrator with the National Aeronautics and Space Administration, testified to a House of Representatives subcommittee on the subject of pilot fatigue in 1999. He summed up the challenge this way. "It has been evident that pilot fatigue is a significant safety issue in aviation. Rather than simply being a mental state that can be willed away or overcome through motivation or discipline, fatigue is rooted in physiological mechanisms...." Therein lies the challenge for most of us. Although many medical doctors are pilots, most pilots are not medical doctors, and the subject of fatigue can be, well, *fatiguing*—for those of us without training in human physiology.

Jay Miller cuts through the complexity and medical terminology to give us a clear and concise handbook on managing pilot fatigue. He begins his book with a vignette from Vietnam and his operational piloting career, but make no mistake, Dr. Miller is one of the world's leading experts on the science of fatigue. More important, the focus of a great deal of his work has been on fatigue *countermeasures*—and that makes his treatment of this subject all the more valuable to those of us who frequently burn the candle at both ends.

This book breaks down fatigue into bite-sized chunks, easily digestible to the average pilot, beginning with a description of the typical error patterns relating directly to alertness and concluding with two appendices that can be directly applied to the challenge. In between, you will learn to come to grips with circadian rhythm, jet lag, and cumulative fatigue. Dr. Miller takes great care to recommend strategies that, if used, can literally save your life.

Like all good books written by pilots for pilots, this one is rich in case studies that allow you to see and learn from the experiences of others. In two decades of trying to come to grips with this difficult subject, this is the single best "how-to" manual I have seen. Take it to bed with you and you will be wiser in the morning. But don't stay up too late.

Fly smart.

Tony Kern

1
Introduction

The crew had been fighting sleepiness throughout this and preceding sorties. They had flown a number of sorties that night, as scheduled by a frag order (fragmentary order) from the 834th Air Division in Saigon. It was business as usual. Now, at about 2 P.M. local time, as the pilot eased 130,000 pounds of C-130E Hercules through 100 feet above ground level (AGL) across the approach end overrun, he controlled the aircraft carefully with his left hand on the yoke. However, his right hand was not on the throttles, where it belonged. Desperately sleepy, he was using it to slap his face gently to help him remain alert during the landing. Not likely, you say? Wouldn't happen? Unfortunately, this is a true story. It occurred in South Vietnam. The year was 1969. I was the pilot.

How did this sorry state of affairs come to pass? It started with crew scheduling practices that failed to account for the basic tenets of human biology. It was continued by the "can-do" attitude that characterizes good aircrews and the management practices induced by that attitude. These factors caused C-130 crews to operate at the very limits of sleepiness during too many in-country missions.

In this small book, I discuss this and other alarming, fatigue-induced incidents experienced by aircrews (private and airline transport pilots and other cockpit and cabin crew members) and examine them in the light of present knowledge about the relevant biology. Through this process, I hope you will learn how to recognize the appearance of fatigue in the cockpit, how to deal with it, and what you can do to minimize the occurrence of aircrew fatigue. I will suggest techniques that might have minimized the effects of fatigue in the incidents described here and that you may use to reduce the likelihood that you will become involved in a fatigue-induced incident or accident. Private and airline transport pilots all share the same, inescapable need for good quality rest, a need that plays as much a role in aviation safety as your levels of training and currency.

Because our concern here is aircrew fatigue, we need to agree on what we mean when we use the word *fatigue*. You have certainly used the word to refer to sleepiness on many occasions. You may also have used it to refer to tired muscles that won't generate the strength you expect or to a tired body after a long run, swim, or bicycle ride. You may have read or heard about chronic fatigue syndrome. Obviously, we're dealing with several different ideas when we use the word *fatigue*. It can mean several different things to one person and several other things to another.

When I talk about fatigue, I refer to a set of symptoms that are known primarily only to you (they are covert) and that depend strongly upon both the task at hand (they are task dependent) and how long you have been awake. Fatigue includes feelings of discomfort or "malaise" that accompany prolonged activity. The fatigue researcher,

I. D. Brown,[1] described fatigue as follows: "Fatigue may be conceptualized as the subjective experience of individuals who are obliged . . . to continue working beyond the point at which they are confident of performing their task efficiently." He defined fatigue as "the subjectively experienced disinclination to continue performing the task at hand." He described "the main effect of fatigue" as " a progressive withdrawal of attention" from the task at hand. Finally, he noted that the latter withdrawal "may be sufficiently insidious that [operators] are unaware of their impaired state and hence in no position to remedy it." I have never found a more cogent description of fatigue than this one, and it certainly fits well with the idea that most of the symptoms of fatigue may be covert.

Fatigue has been on the "top ten" wanted list of the National Transportation Safety Board for more than 10 years now. The board notes that fatigue-related accidents pervade all modes of transportation. To discuss this ubiquitous phenomenon, we need to narrow this wide, abstract concept into more concrete terms. Thus, for purposes of discussion, I divide the development of fatigue into three major categories: acute fatigue, circadian effects, and cumulative fatigue. *Acute fatigue* is the fatigue that develops within one work period and from which you can recover during one major sleep period. The word *acute* is used here in its medical connotation, suggesting a brief occurrence of a condition (for example, one work period).

In my opening example, acute fatigue was induced by several factors. The first was a 12-hour crew duty period (CDP), the time elapsed from the crew's report time until the time the aircraft is in the blocks at the end of the last sortie of the work-plus-flying period or the duty period. The second was a prior period of wakefulness that was about 13.5 hours by the end of the crew duty period. The third was the heat stress experienced during most of our operations in Vietnam. Although the daytime heat stress was extraordinary in the cockpit during much of the year, the first two factors were not extraordinary. Pilots and many others work those hours much of the time. However, just because we work and fly for 12 hours and more or experience CDPs of 12 hours and more all the time without accidents doesn't mean that we *should* do it. The U.S. Congressional Office of Technology Assessment[2] paraphrased a Nuclear Regulatory Commission assessment of the potential advantages and disadvantages of 12-hour shifts compared to 8-hour shifts. They concluded that, because 12 hours of work per day is more fatiguing than 8 hours,

- Alertness and safety may decline.
- Workers may work at a slower pace.
- Workers need more breaks.
- Twelve-hour night shifts are more difficult than 8-hour night shifts.
- Twelve-hour shifts may be difficult for older workers.

My own review of the applied research literature on the subject suggests that the following work factors argue against the use of the 12-hour shift length over the 8-hour shift length:

- Heavy physical work
- Demanding, repetitive mental work
- Safety-sensitive work
- Work requiring sustained alertness (vigilance)

Aircrew tasks incorporate the latter three of these factors. The great emphasis in professional aviation on safety issues seems to have overcome most of the potential risk problems suggested here. However, once we examine error production in business and industry, it should not be surprising to conclude that we work on the edge of accident risk when our workday and flying, or CDP, extend to or beyond 12 hours. The same thing may be said for the private pilot who puts in a full workday before flying, That full workday may include commuting and a working lunch, pushing your preflight to a time that follows 10 hours or more of work-related activities.

Circadian and circasemidian[*] effects occur in three subcategories. First, the two effects combine to produce relatively low mental and physical performance capabilities and extreme sleepiness during the predawn hours, with a similar, but milder impairment during the midafternoon hours. Second, there are circadian effects on sleep quality. The human brain is designed to generate sleep at night. Most often, it does not do as good a job when it tries to generate sleep during the day. The reduced quality and quantity of day sleep makes it much less effective in terms of recovery than night sleep. Finally, of course, circadian biology also affects you strongly when you change time zones (jet lag) and also when you change work shifts from day to night and back.

All three of these circadian-circasemidian effects played a role in my example from Vietnam. Our schedule called for repetitions of this work-rest sequence: five successive work-rest cycles and then a "day off." One work-rest cycle was composed of the 12-hour CDP followed by 14 hours off (the first part of which was spent awaiting transportation and dealing with maintenance). Thus, we existed on a 26-hour (12 + 14) day for almost 5½ days and then had some poorly scheduled time off.

---

[*]Sir-kay'-dee-an: An oscillation with a period of about 1 day, or 24 hours; sir-kah-seh'mee-dee-an: an oscillation with a period of about ½ day, or 12 hours.

Typically, the first cycle started at night, say 10 P.M. We'd get off duty at 10 the next morning and go back on duty 14 hours later, at midnight. The second CDP ended at noon the next day, and we reported back in at 2 the next morning. The third CDP ended at 2 P.M., the fourth at 4 P.M., and the fifth at 6 P.M.

Now, we were given a day off. We'd have some dinner and, being young or relatively young men, we'd generate about 12 hours of recovery sleep. We'd wake up the next morning pretty well refreshed, and we might even get a bit of a nap in the afternoon. Then, after being awake pretty much all day, we'd go back to work at 10 that night to work through the night. By the time that first CDP in the new sequence ended, we'd have been awake for 24 hours, and we'd feel it. We'd be very tired but have little hope of a long, restful sleep period through the middle of the day and early evening.

In terms of circadian biology, the following aspects came into play:

- The combination of the circadian and circasemidian cycles produced impairment during the midafternoon hours, with the impairment peaking at about 3 P.M. I was making my approach just before 2 P.M. Of course, sleepiness during approaches we made in the predawn hours was often just as bad as the event I've described here.

- This was the third day of a 5-day work-rest cycle. I had been awake most of the day before the start of the 5-day sequence, then had worked all night. The schedule then caused my sleep periods to fragment and to occur both in the morning and early in the evening, when the body clock is set for wakefulness, for two more 26-hour cycles. This reduced the hour-for-hour recovery benefit of

the sleep I acquired. Trying to sleep during the day is a problem that aircrews face worldwide, every day.
- The optimal work-rest cycle is 24 hours long. Ours was 26 hours long. Thus, we suffered from a slight jet lag that accumulated at the rate of about 1 hour per day. We experienced the same effect as an aircrew that moves two time zones west each day—you never quite catch up.

Cumulative fatigue is the fatigue that develops across work periods when inadequate rest is obtained between work periods. Circadian biology and sleep disruption both have big influences here. In my Vietnam example, we seldom had the opportunity for a good night of sleep. Consequently, cumulative fatigue built up from CDP to CDP.

How could the inappropriate fatigue I experienced have been avoided? The most beneficial change would have been to put us on regular 24-hour schedules. This would have allowed our circadian rhythms to adjust, eventually, to a single cycle of work and sleep. Now, there is no free lunch. The down side of this suggestion is that many crews would have been assigned permanently to night work. The lucky night crews would work mainly in the period noon to midnight. The unlucky ones would work, for example, from dark to dawn or midnight to noon.

However, even these problems can be mitigated through judicious use of other resources. First, good quality day-sleeping arrangements can be arranged by or for aircrews. Dark, quiet, cool sleeping quarters can be set up, and night workers can conduct housekeeping while the night-flying crews are out. Housekeeping noises in the hallways outside hotel rooms and knocking on your door in the middle of the day are not conducive to good sleep quality or quantity. Second, other groups, such as meal providers and administrators, can also assign night workers to help support commercial crews that must fly at night. Problems such as learning at midnight, just before your show time, that the hotel kitchen closed at 10 P.M. leads to poor nutrition for aircrews. Of course, organizing or finding support resources to operate like this requires management dedication from the top down in an organization. However, if management wishes to reduce the numbers of accidents ascribed to fatigue, they must arrange for adequate aircrew day-sleep facilities.

Although these ideals have seldom been reached in military flight operations, much less in commercial operations, they can be held up as exemplars of how to deal with the problems associated with irregular work schedules and night work. They, along with other suggestions in this book, have the potential to reduce the risks associated with irregular and night flying to a small proportion of their present size. The general aviation pilot may have more flexibility than a large commercial operation in arranging to meet some of the needs of night flying. On the other hand, a commercial aviation organization should be able to exercise greater financial clout in arranging for these needed support activities.

I have aimed this book at one particular audience: aircrews. Thus, I have avoided talking about sleep pathologies such as narcolepsy and sleep apnea and about endocrine pathologies that may disturb the body's daily (circadian) rhythms. If aircrew members suffer from such pathologies, it is at such a low level that the pathologies are not detectable in their required medical exams and do not interfere to any noticeable extent with their flying duties. The reason that I assume the latter for professional pilots is that the extreme feelings of fatigue and sleepiness caused by the combination of long duty days, irregular schedules, and night work plus these pathologies would probably make people drop out of the career field quickly, if they even attempted to enter it. The main problem in this regard may be the development of sleep apnea in aging male pilots. However, because obesity is a factor in sleep apnea severity, and obesity is regulated in the industry, even this pathology may not be a big player.

However, I have one caveat about sleep pathologies for commercial and general aviation aircrew members. You will be the first to detect a problem. It is unlikely that your physician has received training focused on the diagnosis of sleep pathologies. Classes of this nature have been offered in medical school for only about the last decade, and not all medical students take that class. If your level of daytime sleepiness changes for the worse, you may be able to identify the cause of your poor sleep quality and/or quantity: anxiety, musculoskeletal pain, too much caffeine or alcohol, illness and associated medications, and so forth. However, if you cannot identify the source of the problem, a referral to a sleep disorders clinic at a hospital is in order. You must shoulder the responsibility for maximizing your alertness and attentiveness in the cockpit, and flying with an untreated sleep disorder is not an acceptable practice.

The Organization of This Book

It is my wish that, in reading this book, you will become aware of the many complexities that underlie the apparently simple process of acquiring enough rest to perform your job safely and effectively. In addition, I hope that you will acquire some useful knowledge and practices to use when determining how you may best combat the fatigue that comes naturally with the business of being an aircrew member.

The concept of acute fatigue is discussed in detail in Chapter 2. I have called the chapter "It's Been a Rough Day" to reinforce the idea that you don't have to miss a night of sleep or suffer from jet lag to feel, and actually be, tired; all it takes is being awake. Included in this chapter are discussions of the relationship between being awake and becoming sleepy and of the creation of fatigue by the efforts we put forth to meet work demands. A caution is sounded in the chapter about trying to characterize fatigue in terms of gradually diminishing aircrew performance—fatigue develops covertly and may have a sudden, catastrophic effect on performance. Chapter 2 continues with discussions of the idea of task-specific fatigue and of the risks associated with cockpit automation and then concludes with an introduction to the kinds of aviation examples used throughout the book.

The first kind of circadian-circasemidian effect on aircrew performance is discussed in Chapter 3. The chapter title, "The Daily Two-Peak Pattern of Errors," calls attention to the fact that we are at high risk for making errors during the predawn and midafternoon hours. The drawing together of data across many human endeavors to produce a predictive two-peak pattern is described, and there is a discussion of how aircrews may try to deal with this ubiquitous phenomenon.

Chapter 4 provides a very basic introduction to sleep and nap biology. It's not just sleep quantity that can be a problem for aircrews. Information about sleep and nap biology is necessary to understand how sleep quality can be undermined by eating or drinking the wrong things or by trying to sleep during the day—the second circadian-circasemidian fatigue trigger. Also in this chapter, there is a discussion of the excellent work by the National

Aeronautics and Space Administration (NASA) on cockpit napping. Although the Federal Aviation Authority (FAA) does not yet allow cockpit napping, these ideas also apply to napping before flights or trips and during layovers. Strategic nap planning is discussed, also. This kind of planning takes into account sleep inertia (grogginess) that occurs upon awakening from main sleep periods and naps and the effects of changing time zones on the best nap times.

In Chapter 5, I have examined two sleep-related areas of widespread interest. The news media have been involved extensively here, so I have called the chapter "Sleep in the News." The first area of interest has been the question of whether we tend to get enough sleep at night in the United States. Our general sleep behaviors in the population may strongly influence the amount of recovery sleep you plan to get each night between workdays or during periods between trips. They may also influence your expectations about how much sleep you should get on layovers. You will see that, based upon the results of a national poll, we tend to sleep less than we should. The other area of great interest has been the relationship between eating and sleeping. Many magazine and newspaper articles have recommended things to eat to stay awake or to get to sleep. Because many of these recommendations have been based upon very limited amounts of research, I have tried to summarize what we know for each of the food groups, especially in terms of how they may help you sleep or disturb your sleep.

In Chapter 6, I return to the third of the circadian-circasemidian effects on aircrew performance. The phenomenon of jet lag is explored in some depth, along with its differential effects for westward and eastward travel. The natural rates of change experienced by the body clock for both directions of travel are described. I have also explained, with caveats, the latest views on trying to accelerate these rates of change by using melatonin and bright lights.

Much of the previous discussion comes together in Chapter 7, where I have discussed the phenomenon of cumulative fatigue. The contributions to cumulative fatigue by acute fatigue and circadian-circasemidian effects are described here. The value of the siesta, practiced in all of the world except in the industrialized economies, is emphasized for its potential effectiveness in combating the accumulation of fatigue from duty period to duty period. In addition, I have described some insights into the nature, development, and measurement of cumulative fatigue. The role of long-term guilt and anxiety in the development of cumulative fatigue is explored here.

Chapter 8 is dedicated to an update on current views of the main pharmacological approaches used to counter fatigue. I have included both prescription and nonprescription treatments. There are treatments available to keep you alert ("go" pills) and to help you get to sleep ("no-go" pills). Some may be useful for aircrews, but many cautions and caveats are provided.

Finally, in Chapter 9, I have summarized the lessons of the preceding chapters into an aircrew prescription for fighting fatigue. This last chapter also includes a suggested guideline for the reporting of aviation incidents that you suspect were caused by fatigue. Providing this kind of information to such resources as the Aviation Safety Reporting System and to safety investigators will give us a better handle on the degree to which fatigue causes aviation incidents and accidents.

As you scan this book for useful information and practices, please remember that everyone is different. Take the ideas you get here and experiment a little bit to see what might work for you. Of course, you should be sure to experiment at home, when you don't have to deal with work and safe driving. Try altering your sleep behaviors. Try using siestas. Try to find out when your metabolic low-point occurs. Learn about your personal sleep needs. Vary your nutritional pattern to see if you can detect effects on sleep or wakefulness. Learning these things about yourself can only lead to greater alertness, better work performance, and enhanced flying safety.

2

It's Been a Rough Day

The Guantanamo Bay, Cuba, DC-8 crash occurred at about 5:00 P.M. in August 1993. Here is the National Transportstion Safety Board's (NTSB)[3] description of the accident:

> The airplane collided with terrain aprx ¼ mi from the approach end of the runway after the captain lost control of the airplane. Flightcrew had experienced a disruption of circadian rhythms and sleep loss; *had been on duty about 18 hrs* and had flown aprx 9 hrs. Capt did not recognize deteriorating flightpath and airspeed conditions due to preoccupation with locating strobe light on ground. Strobe light, used as visual reference during approach, inoperative; crew not advised. Repeated callouts by the flight engineer stating slow airspeed conditions went unheeded by the capt. Capt initiated turn from base leg to final at airspeed below calculated VREF of 147 kts, and less than 1,000 ft from the

> shoreline, and he allowed bank angles in excess of 50 deg to develop. Stall warning stick shaker activated 7 secs prior to impact, 5 secs before airplane reached stall speed. No evidence to indicate capt attempted to take proper corrective action at the onset of stick shaker . . . . [italics added]

The NTSB noted that the pilots had been awake for about 19 to 23.5 hours. The midafternoon error peak probably played a role here, also—5 P.M. is not too late for such an effect, although the greatest height of the peak occurs at about 3 P.M. The NTSB concluded that the causes included[3]:

> The impaired judgement, decision-making, and flying abilities of the captain and flightcrew due to the effects of fatigue . . . .
>
> Additional factors contributing to the cause were the inadequacy of the flight and duty time regulations applied to 14 CFR, Part 121, Supplemental Air Carrier, International Operations, and the circumstances that resulted in the extended flight/duty hours and fatigue of the flightcrew members.

You don't have to miss a night of sleep or suffer from jet lag to feel, and actually be, tired. All it takes is being awake. On top of that, of course, are the work demands of your 8-, 12-, or 14-hour workday or CDP. The effort you put forth to meet those demands and the costs to your body of those efforts tire you out as well, on top of just being awake. Being awake and responding to work demands add together to lead to the phenomenon we call acute fatigue. *Acute fatigue* is the fatigue that develops within one duty period and from which you can recover during one major sleep period. It develops from wakefulness and work.

Now, if you were an engineer designing a system like a commercial aircraft, and you knew that one component of that system was going to wear out every dozen hours or so, how would you respond in terms of cockpit design? Perhaps you would automate as many cockpit functions as possible to reduce the workload on the cockpit crew. Sounds reasonable. Unfortunately, there are some subtle perils associated with replacing pilots' activities with automation. Automation can lead to accelerated fatigue onset for at least a couple of important kinds of pilot performance: vigilance and decision making.

I have discussed the automation problem below, following a consideration of the effects of wakefulness and work effort on the development of acute fatigue.

The Cost of Being Awake

To the best of our knowledge, the main factor in the development of sleepiness is the prior length of wakefulness. If the analogy of muscle metabolism were to represent brain metabolism adequately, we might expect that intracellular wear and tear, local reductions in available nutrients, and the local presence of metabolic by-products in brain tissue would cause the brain to need some time off for recuperation and that the sleep process fulfils that requirement. In muscles, recuperation includes clearing away and then excreting the by-products of metabolism, regenerating the sodium and potassium concentrations that are essential to muscle cell activity, rebuilding broken-down proteins, and so forth. Perhaps sleep allows the brain tissues to do something like this. It's a nice analogy but, in truth, we simply do not know why sleepiness follows prolonged wakefulness.

However, we do know that we can predict subjective reports of sleepiness and subsequent sleep amounts reasonably well with a straight-line representation of depletion during wakefulness. A number of quantitative models exist these days for the prediction of human performance as it is affected by wakefulness, sleep, and daily (circadian) rhythms. For my descriptions, I have depended upon the model by Dr. Steven Hursh, now with Johns Hopkins University and Science Applications International (SAIC), and incorporated into the Windows program called the Fatigue Avoidance Scheduling Tool (FAST), developed under Air Force contract by NTI, Inc., of Dayton, Ohio. The depletion function, caused by wakefulness, is simply a straight line representing decreasing pilot performance effectiveness across the waking period. In terms of the percentage of maximum pilot performance on mental tasks for a well-rested pilot who slept from 11 P.M. to 7 A.M. the previous night, the straight line declines from about 98 percent performance effectiveness in the morning to about 86 percent performance effectiveness by bedtime.

Once we reach the 86 percent level, we are about ready for another night of sleep. Of course, we often allow ourselves to fall well below that level. Without any countermeasures, your nighttime performance effectiveness will fall to about 80 percent at about 3:30 A.M., a natural, predawn performance trough. With sleep loss, you can drop down to a low point of 60 percent performance effectiveness instead of 80 percent. In my experience, I worry about the safety of pilots who are projected to fall below 90 percent.

Work, Effort, and Fatigue

The concept that fatigue produces a decrement in work performance is not new. As early as 1884, Mosso examined the association between muscular fatigue and work decrement. In industry, during the years before the Second World War, little regard was given to the individual in the workplace. Any concern about worker fatigue focused on its effect on production. However, with the rise in importance of air power in World War II, military fatigue studies at Wright Field in Dayton, Ohio, and Randolph Field in San Antonio, Texas, shifted toward emphases on mental processing and psychomotor tasks, functions used extensively by pilots. These studies continue today, with focal points including the Air Force Research Laboratory's Warfighter Fatigue Countermeasures Research and Development Group at Brooks Air Force Base in San Antonio, the FAA Civil Aeromedical Institute in Oklahoma City, Oklahoma, and the Human Factors Division of the NASA-Ames Research Center in Mountain View, California.

The *work demanded* of an aircrew member is viewed as a stress, to which you respond with some evidence of stress and strain. We differentiate physical (muscular) demands from mental, or cognitive, demands. The *requirements* to help load or unload cargo or to spend too much time in an extremely hot cockpit on the ramp are examples of physical work stresses. The *requirement* to fly a safe instrument approach with the weather at minimums is an example of a cognitive work stress.

An example of *strain* in the physical domain would be the metabolic *effort* required to help load or unload the cargo. An example of strain in the cognitive domain would be the cognitive *effort* required to fly the safe instrument approach with the weather at minimums. The degree of effort you bring to bear on a specific work demand may act to reduce the work demand, depending upon the task. If you or your taskmasters are satisfied with the results of your effort, you have reduced the work demand to zero for the moment. If satisfaction has not been achieved, you must decide whether you will bring more effort to bear on the demand. Your level of motivation modulates the total degree of effort you decide to put forth. Specifically, greater motivation leads to greater efforts. In addition, a stress response within the human body may be a mediator or pathway between external work demands and metabolic strain. For example, when physical or psychological stresses depress immune responses, we call the depression a stress response. Such a stress response may help trigger a strain such as catching a cold.

There are a number of *physiological costs* associated with physical effort. Physiological costs are metabolic in nature and may include the following, among others:

- Elevated whole-body metabolism associated with nonsedentary work, like jogging
- High levels of specific muscle anaerobic metabolism associated with lifting or with the maintenance of a single posture for a long time
- Relatively high heart muscle (myocardial) metabolic demands due to the combination of poor physical conditioning and high physical workloads
- Increased potential for the triggering of central nervous system sleep systems (falling asleep on the job) associated with sleep disruption

Similarly, there are *psychological costs* associated with effort. These may include the following, among others:

- Loss of motivation
- Feelings of anxiety

- Boredom
- Loss of the ability to remain vigilant

Performance is often the "bottom line" when we are concerned about aircrew fatigue. We are often asked whether a crew duty period or flight may be accomplished safely. This is a reasonable question. Most often, the answer is Most of the time. After all, aircrews make many, many flights in very fatigued states and do not have accidents. However, one's overt performance is not always sensitive to the effects of fatigue. This problem is due to the "two-edged sword" of human adaptability. The "good" edge is the ability of aircrews to motivate themselves to face challenges and to accomplish difficult tasks in acceptable manners in the presence of high levels of strain and resulting fatigue. Typically, the fatigued but motivated human can mobilize his or her physical and cognitive resources quite well for brief periods. This is the "can-do" attitude, characterizing what we think of as good aircrews.

The "bad" edge of the sword is the eventual effect of physiological and mental costs: There may be a cessation in performance (a mental lapse) or an involuntary onset of sleep (falling asleep on the job). Thus, the measured performance of the fatigued but motivated crew member may show no impairment at all until performance ceases abruptly. Thus, I view *fatigue* as a covert result of the costs generated by effort and performance. I seek evidence of fatigue in the perceptions of crewmembers, in levels and variability of performance that are diminished subtly below reasonable expectations, and in behaviors associated with sleepiness.

Fatigue may also lead to injury. An acute physical stress that exceeds connective-tissue limits may lead to a sprain or strain of a joint. Lesser physical stresses, repeated for days, months, or years, may cause cumulative or repetitive stress injuries. Sedentary work in the absence of exercise and nutritional limitations may lead to morbid obesity and to cumulative trauma of the back. Excessive aerobic effort, especially in a hot environment like a hot aircraft on the ramp, may lead to heat exhaustion and to myocardial ischemia, raising the possibility of heart muscle damage. The impairment of cognitive abilities may lead to poor decision making and risk taking behaviors and subsequent incidents. Thus, evidence of fatigue presence is often seen in increasing numbers of otherwise unexplained personnel injuries, real property damage, and close calls.

This description of fatigue certainly applies to the causes of the general malaise we feel at the end of a long workday or CDP. In addition, we should ponder one kind of task-specific fatigue. I am concerned here primarily with the psychological costs caused by mental effort. Certainly, there are physiological costs associated with long flights: specific static muscle fatigue from sitting in the cockpit, dynamic muscle fatigue for flight attendants, joint stiffness from sitting, dehydration, and so forth (long-haul syndrome for cockpit crew and passengers). However, cockpit crews face one particularly difficult mental task: monitoring cockpit automation. This is the most likely cockpit task in which performance may cease because of a mental lapse or an involuntary onset of sleep (falling asleep on the job). The cockpit automation that can cause problems includes everything from a simple wing leveler to autopilots that navigate and land the aircraft.

Task-Dependent Fatigue

Because the name, task-dependent fatigue, may be new to you, here is an example to help familiarize you with the idea. First, we need to set the scene: Although you usually sleep well, about 7.5 hours per night, and have done so lately, you were able to acquire only about 3 hours last night. Now, it's 2 P.M. and you "feel" very sleepy. Someone would have to know you quite well to detect this sleepiness by just looking at your face, and other drivers on the highway certainly can't detect your sleepiness easily; perhaps you are weaving in the lane a little more than usual, but can they detect that? Your sleepiness is covert. Others cannot detect it easily unless you tell them that you are sleepy.

Your task in this example is to continue driving on Interstate 70 across Kansas, as you have been doing for several hours, which is boring at best. You are quite likely to fall asleep at the wheel. Now, let's change your task to one that is more enjoyable. Your alternative task is to keep reading a novel by your favorite author. You may feel sleepy later today, but right now you feel great. Your sleepiness is task dependent. When the task required that you remain alert in a boring situation, you felt very sleepy. Now that the task requires a more enjoyable activity, you do not feel sleepy at all.

Some folks may argue that task-dependent fatigue is really a person's habituation to a task and is not, truly, fatigue. They might say that the task is simply boring and that this is not fatigue. Whatever you call it, the fact is that your task performance declines as your time performing the task continues. This is the same overall pattern that we have with wakefulness. Your performance declines as your time awake continues. These two effects can be additive.

If you cannot change tasks, then, much like the effects of biology, the effects of task-specific fatigue can't be avoided. They can only be recognized and managed. Fortunately, it seems as though aircrews do not suffer very much from task-specific fatigue. They may suffer from sleepiness and reduced levels of vigilance that affect cockpit task performance, but their focus on the overall task of flying seems to remain intact. Whether this is due to motivation, automated behaviors, and/or other factors is not known.

Cockpit Automation

When I teach human factors engineering at the introductory level, there is one point I always emphasize strongly: the need to consider human variability in system design. When a system is designed such that a human operator is to be used, the most variable, least predictable component of the system will be the human operator. You and your aircraft are a human-machine system. If you don't have a death wish, you always make sure to the best of your technical and financial abilities that the mechanical components of the system are up to date in terms of inspections, repairs, replacement, and updates. Because of the excellence of engineering in the aviation industry, the mechanical components in your aircraft serve you well. Most likely, you have never had a catastrophic mechanical failure in your aviation career (such as in-flight engine loss in a single-engine aircraft); accident statistics support this contention.

Now, consider the human component in this system. The quality of your contribution to system performance varies widely as functions of time of day and prior time spent awake. However, the design of the cockpit and the design of the airway system do not adapt themselves to the fluctuating capabilities of the pilot. There is an implicit assumption in system design that your performance will be constant at all times. You experience the effects of this design philosophy when

It's Been a Rough Day 29

you try to drive while you are very sleepy. Either your lateral lane position will vary more widely than usual (weaving), you will find your speed varying more than you wish above and below your desired speed, or both. (These common manifestations of drowsy driving have been quantified a number of times in open highway and simulator driving research studies.[*]) The designs of the automobile and our highway systems do not adapt to your performance capability. Enter the cruise control.

The cruise control relieves the static muscle and tendon stress caused by continuous pressure on the accelerator pedal; automates the simple longitudinal highway tracking task we call speed control, allowing you to spend more time looking out the window (pilots appreciate more than others the value of that freedom); and allows you to maintain a consistent highway speed even when you are sleepy. The cruise control is a good example of how design engineers may use simple, straightforward automation to reduce the variability of human input to a human-machine system.

Although system designers have used automation to deal (sometimes well, sometimes not so well, as discussed below) with the variabilities of human inputs to the system, the accident investigation and reporting systems, and their lessons-learned components have failed to keep pace. In the late 1980s, the U.S. Air Force realized that even though its rate of Class A mishaps (fatality, loss of aircraft, or major damage) had fallen to only about one per 100,000 hours of total flying time, about

---

[*]In fact, I was present at the invention of the United States' version of the Lane Tracker, a tool for measuring weaving for open-highway research studies. Reported in Mackie and Miller.[4]

80 percent of those accidents were caused by human errors. It was suspected that the causes of many of these accidents could be traced to inattention or distraction during maintenance activities or while airborne. I contended that lack of adequate sleep and the effects of the two-peak daily error pattern caused a large proportion of this inattention and distraction.

Of course, saying this is easier than quantifying it. In fact, historically there has been little thought given to the idea that night flying may lead to an inordinate amount of time spent awake, cause flight legs and approaches to occur during the predawn trough in human metabolism, and take the place of needed sleep. For example, in the Aircraft Owners and Pilots Association's 1999 Nall Report, the dangers of night flying were characterized solely in terms of its interaction with weather and instrument approaches. There were no conclusions drawn relating to fatigue effects. I suspect that this omission is due to a general lack of useful data in accident investigation reports. However, in the last decade, accident investigators have delved more and more into the 72-hour work-rest histories of accident aircrews, trying to determine whether fatigue may have contributed to the accident. The fact that I've been able to cite a number of accident examples in this book shows that progress has been made in terms of understanding the contributions of fatigue to accident incidence. Now, back to automation.

Automation design should exploit the strengths of the pilot while protecting the aircraft systems from the pilot's weaknesses. Two human strengths that may be exploited in design are visual pattern recognition and supervisory control. One human weakness that must be dealt with is the general inability of the human to sustain attention in low mental workload conditions (to remain vigilant).

The way in which we selectively allocate aircraft system functions to the human operator or to automation has received quite a bit of attention in aviation research. The most common philosophy applied to automating a control system is to automate repetitive, manually controlled processes (such as holding the wings level and holding a heading or an altitude), increasing the freedom of the human operator to engage in planning, goal selection, pattern recognition, fault detection, and other cognitive, supervisory functions. This rationale is reasonable because manual control, particularly over an extended time, is not a particular strength of the human operator. Given adequate feedback, an automated system can perform tedious and complex manual control tasks far better than a human can. On the other hand, this automation philosophy often forces the human into the role of a system monitor, a function handled poorly in most cases by the human brain and which sets up perfect conditions for lapses in attention.

Concerns about cockpit automation include

- Increased monitoring load placed on the pilot.
- High level of responsibility with little to do.
- Loss of manual skill proficiency.
- Pilot-out-of-the-loop problems.
- The human's uniquely elegant pattern recognition abilities being replaced by less-competent sensors.
- The classic vigilance problem characterized in the story of the boy who cried, "Wolf!" – that is, a high false alarm rate will cause the pilot to ignore indications of system malfunction, and malfunction detection rates by the pilot will drop to near zero.

- Displays not supporting optimal performance by the pilot, and not allowing for variations between individuals in the ability to remain vigilant.
- Misinterpreting automation problems that are caused by pilot vigilance failures and the wrong "fixes" being implemented.
- Subtle vigilance problems not being recognized until after the occurrence of many accidents involving automated aircraft.

Lapses

Dr. Raja Parasuraman and his colleagues at George Washington University noted that the coding manual of the NASA-Ames Aviation Safety Reporting System (maintained primarily for the FAA) defined complacency as "self-satisfaction which may result in nonvigilance based on an unjustified assumption of satisfactory system state."[5] They noted that complacency includes failures to respond to an automation malfunction.

With fatigue comes the increased likelihood of periods during which we fail to respond to important occurrences in the low work load environment. These periods may be brief "lapses." During a lapse, we are usually able to carry on highly learned, automated behaviors in unremarkable environments. For example, most of us have experienced an occurrence of a lapse while driving. We realize suddenly that we cannot recall any information about the last mile or two we've driven on the highway, but we have negotiated that part of the highway without an accident.

The situations associated with the occurrence of lapses include the absolute and relative rates at which important and unimportant occurrences occur in the environment; the complexity, timing, and conspicuousness of the occurrences; the sense involved (vision, hearing); the time of day; the total work load placed on the individual; the work schedule and its interactions with sleep quality and with bimodal circadian rhythms in human performance; the individual's motivation; attention from management; and others. That's pretty complicated. The bottom line is that the occurrence of lapses in the cockpit or before flying should signal to you that you are sleepy and less vigilant than you should be.

Supervisory Control

Dr. Neville Moray, respected for his research in this and related areas, notes that among the supervisory functions allocated to the pilot, perhaps the most critical is the combination of failure detection and fault diagnosis. How do we accomplish this feat? The work of Dr. Jens Rasmussen, well known for his studies of mental processes, suggests that the human cognitive, supervisory functions of interest for failure detection and fault diagnosis may be those characterized as knowledge based, as opposed to those functions that are skill or rule based. Knowledge-based functions include the recognition of, and creation of responses to, novel and unpredicted situations for which clearly applicable rules do not exist. This capability is, perhaps, the greatest strength that a pilot can bring to the cockpit. It is a capability that has eluded programmers of computers, at least to date. Computer programs can implement skill- and rule-based operations quite well. They do not apply knowledge-based functions well in the cockpit.

Aviation research that deals with analytical behaviors suggests that the human operator in an automated cockpit will, in most cases, be forced to diagnose system failures under the pressure of time stress. This is due, simply, to the absolute speed at which aircraft operate and the speed with which they move relative to each other. Instead of forming hypotheses in a relatively leisurely manner, pilots probably use a pattern-matching approach to failure diagnosis. Is this function impaired by fatigue? It appears that it is.

In a study I conducted to determine fatigue levels experienced by Coast Guard cutter crews, I found several lines of evidence that suggested the development of mild cumulative fatigue.[6] When I examined the day-to-day trend in pattern-matching performance (measured with a specialized, computerized instrument), I found that pattern-matching performance decreased from day to day for the crew members I studied on three of the five cutters in the study. With any practice effects, one would expect performance to increase from day to day. Thus, one may conclude that pattern-matching performance is susceptible to the effects of fatigue. My finding was not unique. Other investigators have found similar effects of fatigue on pattern-matching performance.

Examples

Federal regulations limit pilots in commercial operations to 8 hours of flying and require 8 hours of rest each 24 hours. These numbers, by themselves, seem reasonable solely in terms of the management of acute fatigue, although the question of how one acquires 8 hours of sleep during only 8 hours of rest remains a mystery to me. However, many liberal interpretations have been made of these numbers. For example, one might reset the 8-hour counter at midnight so that 16 continuous hours of flying would be possible, in theory. Without regulations concerning duty time limits (CDP), airline industry contractual negotiations have led to the adoption of CDPs of 14 to 16 hours per 24-hour period. Some research suggests that a 12-hour CDP might be the outside limit in terms of acute fatigue issues.

One accident that seems to support this 12-hour acute fatigue limit was the June 1999 American Airlines accident in Little Rock, Arkansas, which killed 11 people. It was the first day of a three-day sequence for the crew. Thus, one would hope that cumulative fatigue was not a factor. The accident occurred during a landing, in heavy wind and rain, at about 11:40 P.M. Thus, predawn and midafternoon performance effectiveness problems were not players. The pilots had reported for duty about 10:15 A.M. At the time of the crash, they were about 13.5 hours into their 14-hour CDP. The NTSB concluded that fatigue was a factor in this accident, questioning the decision to make the approach in such difficult environmental conditions and noting the pilots' failure to deploy the aircraft's spoilers on landing. If I were asked to categorize the kind of fatigue that was a factor in the apparently impaired decision process here, I'd lean toward acute fatigue.

For this book, I extracted several aircraft accident examples from the files of the NTSB to illustrate the various effects of fatigue on pilot behavior and performance.[*] In reading the examples, you need to remember that fatigue-related accidents are almost always caused by a combination of fatigue factors, including acute and cumulative fatigue, sleep disruption, and circadian and circasemidian rhythms. The long duty day that ends in an

---

[*] I had hoped to use examples from the Aviation Safety Reporting System (ASRS), operated at the NASA-Ames Research Center. However, to help provide anonymity in reports, times of occurrence of incidents, and the crew activities related to the incidents listed in that database are masked by the NASA database managers. Thus, it is virtually impossible to distinguish among acute and cumulative fatigue and circadian factors in those reports.

accident is usually accompanied by truncated sleep at the beginning of the duty day or a flight that lasts into the late hours of the night, when sleeping would be the biological expectation for the brain and body. If the flight extends into the predawn hours, the natural, predawn trough in performance effectiveness comes into play as an additive factor. Similarly, normal acute fatigue without adequate recovery between duty periods leads to cumulative fatigue. Accidents in which acute and cumulative fatigue are obvious contributors often occur during the predawn trough and occasionally during the natural, midafternoon slump in performance effectiveness.

Accident investigators are quite familiar with the idea that accidents are caused by multiple factors. In fact, one reason that major accidents are rare is the idea of joint probability. For example, in the fatigue arena, the probability that the pilot's capabilities are severely impaired must occur coincidentally with a system or environmental problem that needs the pilot's attention. If the independent probability of each of these occurrences is small, the probability that both will occur at the same time is even smaller. This is true because the probability of the joint occurrence requires the multiplication of the two probabilities. For example, if each probability is about 10 percent, the joint probability is about 1 percent (0.10 × 0.10).

This joint probability phenomenon characterizes the following accident example that occurred on October 2, 1990 in New Hampshire[7]:

> The airplane [a Cessna 172P] was being ferried to Europe and was on a great circle route at night between Syracuse, NY, and Bangor, ME. The plane was in cruise flight at 5500 ft MSL [mean sea level] when it struck Mt. Washington. It was equipped with an autopilot and LORAN C.

According to the LORAN manufacturer, the airplane was within ½ mile of course and on heading at the time of the accident. The top of Mt. Washington was obscured. However, a witness 7 miles away saw the lights of the airplane and a fireball from the impact. *The pilot had been awake for 22 hours when the accident occurred* [italics added].

It seems obvious from our previous discussion that several fatigue-related factors came into play here. The first is acute fatigue: The pilot had been awake for 22 hours. Second, perhaps there was a lapse in vigilance while the autopilot controlled the aircraft. Finally, the predawn trough in attention and performance played a factor: The accident occurred at 4:50 A.M., local daylight savings time.

These things happen to private pilots, as well. For example, this accident occurred about a half an hour after midnight on June 28, 1998[8] (described in part in Tony Kern's foreword to this book):

The non-instrument rated private pilot (237 hours of flight experience) was a professional truck driver and had worked 12 hours on the day of the accident. The pilot departed at 2200 (he had 5.2 hours of night experience during the last 18 months). The moon set at 2255. Witnesses reported seeing a large spark which lit up the sky on the approach end of runway 4. The power went out at the airport, including the runway lights. N7366M was seen flying low over the runway and applying power to go-around. The airplane was found in a cornfield 1,500 feet left of runway centerline. The approach track to runway 4 crosses transmission lines 3,510 feet from the PAPI indicated runway touchdown point (the three top lines were found separated). It could not be determined if the pilot had training or exposure to the PAPI system. The pilot last flew with his flight instructor (in N7366M) on March 19, 1998. The flight instructor observed that the attitude indicator was not functioning properly. No documentation was found which indicated that the attitude indicator had been fixed.

The NTSB concluded that the probable cause was[8]

> The pilot's failure to maintain adequate obstacle clearance and the proper landing glide path. Also causal was his loss of aircraft control during an attempted go-around. Factors were the dark night light conditions, the pilot's recent lack of sleep, and the pilot's lack of recent night flying experience.

In this case, after working for 12 hours, the pilot made a questionable decision to fly and then augured in after at least 14.5 hours (12 + 2.5) of combined work and flying.

The lesson to be learned from this discussion of acute fatigue is that you cannot avoid the occurrence of some degree of fatigue during your workday or crew duty period. The longer you are awake, the sleepier you will become. The more mental effort and physical effort you put forth, the greater the cost to your metabolism. These are natural phenomena associated with human biology. If your workday or CDP were only about 8 hours, the fatigue you develop in one workday or CDP would hardly ever be a problem in terms of your work performance. However, at some point after 8 hours, the probability that you will make errors starts rising disproportionately as time continues to pass.

How should you deal with these phenomena? First, be fully rested when you fly or start a trip—carry no cumulative fatigue into the flight or that first CDP; it will simply add to the acute fatigue you experience. Thus, when preparing for a flight or for a trip or when on a trip, you should trade awake time for sleep time whenever possible and legal. Second, you will learn here about sleep biology and circadian rhythms and how they affect acute fatigue. You will also learn how to plan naps and to redesign your sleeping habits and accommodations. Use this knowledge to help combat acute fatigue.

3

The Daily Two-Peak Pattern of Errors

We are all aware of daily ups and downs in our abilities. Most of us complain of the "postlunch dip," characterized by sleepiness and low motivation. Such normal fluctuations in performance have been documented for many years and are not random throughout the day and night. The mistakes[*] and slips[†] that lead to errors of omission and commission and then to aviation incidents and accidents are most likely to occur during specific portions of the day-night cycle: the predawn and midafternoon periods. Knowing this, you may be better able to schedule some aviation activities to avoid these periods.

We'll never know from the following NTSB report whether this accident was caused by an error of omission or an error of commission, but it certainly appears that the predawn trough in attention and performance played a role. The accident occurred in mid-November at 4:36 A.M.[9]

[*]Mistake: Forming the wrong intention.
[†]Slip: A failure to carry out the correct intention.

The pilot was on his last flight leg for that evening carrying cancelled bank checks. He was cleared for an instrument landing system (ILS) approach to runway 02 to the Springfield-Branson regional airport by Kansas City Center. The aircraft crashed about a mile short of runway 02 while on the night instrument approach. The weather at the time of the accident was reported as two hundred feet overcast with visibility at one and a quarter mile in mist. Winds were reported at one five zero degrees at nineteen with wind gusts to two four. Altimeter setting was 30.24 inches. The Kollsman window of the altimeter in the aircraft was found after the accident set to 30.50 inches. Kansas City Center transmitted to the pilot "the new Springfield weather just came out uh has still has two hundred feet overcast visibility uh one and one-quarter mile now and uh mist wind one five zero at one niner gusting to two four altimeter uh is uh three zero two four." The pilot acknowledged "three zero two four, Prompt Air five fifty." The baggage handler, who loaded the airplane before the pilot departed for Springfield, said that the pilot "looked very tired and fatigued." The pilot had commenced his workday at approximately 1800 CST the day before the accident.

Numerous studies have demonstrated that human circadian and circasemidian rhythms actually cause a *two-peak* daily pattern in incidents and accidents.[*] As early as 1955, Bjerner reported on the daily performance fluctuations for employees at a Swedish gas company whose work was monitored over a 20-year period. In

---

[*]This and other technical references in this chapter may be found in Mitler and Miller.[10]

the distribution throughout the 24-hour day of 74,927 meter reading errors made by these workers, more errors occurred during the night, with a major peak between 1 and 3 A.M. A smaller afternoon peak in errors occurred between 1 and 3 P.M.

Since these early observations, other more destructive human error events have also been shown to occur with this same two-peak pattern. For example, there was the 24-hour distribution of 6052 vehicle crashes attributable to fatigue, crashes for which investigations disclosed no mechanical failure and no alcohol or substance-related causal factors. A two-peak pattern was apparent in this distribution, and its timing was similar to that of the gas meter reading errors. The number of crashes was elevated between about midnight and 6 A.M. and again between about 1 and 4 P.M.

This two-peak pattern is not restricted to behavioral measures. Very similar patterns also exist in biological measures that seem relatively unrelated to behavior. One is the timing of human mortality attributable to disease and susceptibility to toxins. Another is sleep tendency. With the application of continuous brain wave (electroencephalographic, or EEG) monitoring techniques for the measurement of cycles in sleep and wakefulness, it became apparent that sleep tendency also has its ups and downs throughout the 24-hour day. Exactly why increases in the biological tendency to fall asleep are associated in time with increased mortality and susceptibility is not known. However, one leading hypothesis is that the waxing and waning in both sleepiness and susceptibility are controlled by the same or similar biological clocks. The timing of these circadian peaks and troughs seems to be related to the relationship the human circadian system has to the daily light-dark cycle. The circadian patterns of peaks and troughs seem to be only distorted, but not shifted in time, in people who work at night and sleep in the day. Thus, the timing of these peaks and troughs in performance is probably not under a great deal of voluntary control.

To find commonality, the two-peak patterns found in large databases for traffic accidents, human mortality, driver drowsiness, operators' delays in answering calls, locomotive auto-brakings, and meter reading errors were combined. The result was a generalized, two-peak pattern across the 24-hour day in which there was an extremely sharp error peak at about 3 A.M. and an error peak about one-quarter that size at about 3 P.M.

These powerful data show unequivocally that there are critical periods before dawn and in the middle of the afternoon during which sleepiness and inattention may lead to mistakes and slips that cause incidents and accidents. This phenomenon is biologically based, stemming from the interactions of circadian and circasemidian rhythms, and it clearly affects a variety of activities, including aviation.

Conversely, the low points of the two-peak pattern illustrate for us those times during the day when it is difficult to sleep. Generally, it is difficult for the brain to generate sleep during the dawn to noon period and the late-afternoon to late-evening period. This information will come in handy, later, as we consider when to sleep during trips.

Here's an accident involving an error of commission that illustrates how the midafternoon slump in attention and performance can bite you. In this case, cumulative fatigue also played a role. The accident occurred at about 2:45 P.M. to a student helicopter pilot executing a hover taxi.[11]

While returning to the ramp, flying about 4–5 feet above ground level, the student pilot noticed that the main rotor RPM was decreasing. He lowered the collective and added power but the rotor RPM was not changed before impact with the ground—first on the rear portion of the skids. The helicopter rolled onto its side and both occupants exited. The pilot further stated that there was no engine failure or malfunction, and contributing to the accident was the fact that he had only obtained 2–3 hours rest each of the past 4 nights.

Finding Your Predawn Low Point

You can figure out the approximate time at which your metabolism hits its low point. It is the time at which you feel the worst during the predawn hours and are most likely to commit errors. Here are two tools to try to find this point. To use either tool, you must first be at home for enough days to feel that you are well established on your normal, desired home schedule. You then record your own observations on the first night that you must work after this period at home. You make your records in terms of home time, no matter what time zone you are in for work. Make records for just the first night of work—after that, your circadian rhythm will start to change. However, make the records on at least three first nights so that you can choose a relatively reliable estimate. Data from just one first night may be flawed by some factor I haven't considered here, such as illness, anxiety, and so forth. Engineers tell me that you should always measure things once or three or more times. You can measure it once and believe it (a bit risky), or you can measure it several times and take an average. If you measure something twice and the two values don't agree, you don't know what conclusion to draw.

The first tool you could use to find your personal low point is the Stanford Sleepiness Scale (SSS). You would use the SSS to track how sleepy you feel during the night. It should be sensitive enough to provide the information you need. The SSS was published in the open research literature in 1973 by E. Hoddes and colleagues at Stanford University. It has been used extensively since then in research and medicine to track sleepiness. It looks like this[12]:

1. Feeling active and vital; alert; wide awake
2. Functioning at a high level, but not at peak; able to concentrate
3. Relaxed; awake; not at full alertness; responsive
4. A little foggy; not at peak; let down
5. Fogginess; beginning to lose interest in remaining awake; slowed down
6. Sleepiness; prefer to be lying down; fighting sleep; woozy
7. Almost in reverie; sleep onset soon; lost struggle to remain awake

Copy this scale onto a card and keep it handy. For example, carry it with you in a pocket-sized appointment book. Then select a number, 1 through 7, that best describes how you feel each hour during the night. The hour with the highest number (greatest sleepiness) will be your physical and mental low point for your normal waking and sleeping schedule at home.

Body temperature is another tool you can use to find your low point. That low point in performance is also usually about the time that your body temperature reaches its lowest point during the predawn hours. Thus, just as above, when you have been at home for a number of days and feel that you are well entrained to your normal, desired schedule, keep track of your oral temperature each hour on the first night that you have to work. To obtain a useful oral temperature, do not eat or drink for the 15 minutes before you take your temperature; food and drink will change the local temperature in the mouth so that it does not represent body temperature. The hour with the lowest temperature will be your physical and mental low point for your normal waking and sleeping schedule at home.

Using the Knowledge

How can you use your knowledge of the daily two-peak pattern in errors, incidents, and accidents to improve aviation safety? First, for those of you in commercial aviation, you should respect the huge increase that has occurred within the last century in the amount of energy a pilot controls. An airline pilot in 2001 controls several orders of magnitude more potential and kinetic energy than a mail pilot of the 1920s. On the job, when the pilot is sleepy, there is much more at risk than in earlier years. However, the biology and sleep needs of both pilots are identical. In some ways, we are primitive beings who operate very demanding systems, the designs of which do not account for biology.

Second, you should recognize that we have not yet designed very many automated commercial aircraft that use human operators effectively. Any such designs should exploit natural human strengths such as pattern recognition and decision making. The designs should also protect the system from natural human weaknesses such as vulnerability to attentional lapses in boring surroundings, especially during critical periods of the 24-hour day and under conditions of sleep deprivation. Perhaps the best cockpit automation designs are the simpler manual control features found in general aviation aircraft. They allow the pilot to apply natural human strengths to the progress of the flight and reduce the contributions made by continuous manual control to pilot fatigue.

Third, you should recognize that in the modern era it is impossible to avoid assigning a significant portion of the commercial aviation work force to night work. Cockpits, cabins, maintenance facilities, towers, and air route traffic control facilities require around-the-clock staffing. In conjunction with this awareness, you should recognize that night work always compromises a person's ability to acquire adequate sleep. The private pilot who expects to work all day and fly all night is a great risk to all.

One corrective approach that applies here is the effort that has been made by NASA-Ames to specify structured cockpit napping for the crews of civil and military transoceanic flights who must deal with extreme jet lag and/or long duty days. The military presently leads commercial aviation in implementing this correction. At present, this is an unlikely option for general aviation.

Fourth, you should recognize that the methods for identifying unacceptable sleepiness that are currently available to laboratory researchers do not lend themselves to widespread use in the cockpit—even if we could agree on some specific level of sleepiness as representing unacceptable job performance. However, a number of laboratory-originated performance tests have been adapted for daily workplace use and could be made available for preflight self-assessments by pilots.[13]

Once you accept these four ideas, what can you do with our current information and technology? You now know the periods within the 24-hour day during which mistakes and slips leading to incidents and accidents are most likely to occur. Therefore, it may be possible for you or others to adapt flight schedules appropriately, integrating fundamental principals of sleep and circadian rhythm biology. For example, you may

- Determine those flight legs that are most vulnerable to loss of attention. Some of these legs require higher levels of vigilance than others. Conversely, other legs require time-sharing of attention among several dynamic cues (very high mental workload). Pilots operating in areas with high traffic density fall into this category. Instructional flights may also belong here.

- Examine the recent patterns of incidents (mistakes and slips; errors of commission and omission; judgment errors) that occur in these or similar legs.

- Rank these legs using a scale or set of scales that reflects each leg's vulnerability to errors and the potential for loss of life, health, and property if an attentional lapse occurs. The scale(s) might be devised by the pilot, instructors, check crews, or safety or quality circles, any group that observes flight safety with a critical eye. There may be more than one dimension used to characterize various legs. Examples might include how far the job deviates from moderate mental workload toward either high or low (boring) mental workload, how likely an error is to injure someone, and so forth.

- Whenever practical, avoid scheduling critical flight leg flights during the predawn hours.

- When flight legs must be accomplished during the predawn hours, plan to manage the elevated risks associated with flight during those hours. Assume that you will make a much higher number of errors of omission and commission than usual during that period. Have all crew members double- and triple-check each other's work. For private pilots, bring along a copilot qualified to double-check your work.

- When flight legs must be accomplished during the midafternoon hours, plan to manage the elevated risks associated with flight during those hours. Assume that you will make a higher number of errors of omission and commission than usual during that period. Have all crew members double-check each other's work. For private pilots, bring along a copilot qualified to double-check your work.

- Introduce preplanned naps into the cockpit, when legal and practical.

- Educate yourself about the intelligent management of your principal sleep periods and naps. Most people view sleep as a passive state of the brain. It is not. The brain generates sleep actively, and it needs the proper environmental conditions and time of day to do that efficiently. Simply educating yourself about what conditions the brain needs to generate good sleep can help. Brochures are available from many sources to help with this effort.

- Use self-help fitness-for-duty testing so that you may understand your personal impairment from sleep deprivation. As with alcohol, we tend to overestimate our competence when we are fatigued. If you are a well-practiced computer-game player, make a systematic study of your scores as they are affected by time of day and night and lack of sleep.

Although these steps can be effective in combating the natural error potential peaks that occur each day, they can also be effective in combating aircrew fatigue in general. You'll see these ideas again in this book's overall prescription for fighting fatigue.

4

Sleep Biology and Napping

> Q. *What is sleep?*
> A. *How can you know sleep when you are awake? The answer is to go to sleep and find out what it is.*
> Q. *But I cannot know it in this way.*
> A. *The question must be raised in sleep.*
> Q. *But I cannot raise the question then.*
> A. *So, that is sleep.*
>
> —Sri Ramana Maharishi

When we suffer from fatigue and sleep loss, we tend to shed tasks other than those that we know we must face. We develop a lack of initiative. Here's an example, abridged from an NTSB report[14]:

> At approximately 9:47 P.M. PST, a DC-8 on a 14 CFR 121 nonscheduled international cargo

flight from Seattle-Tacoma International Airport, to Anchorage, Alaska, lost its number 1 and 2 engine cowlings on takeoff from Seattle. Following the separation of the number 1 and 2 cowlings, the flight returned to Seattle-Tacoma International and landed without further incident. However, post-accident inspection of the aircraft revealed substantial damage to the aircraft's left wing and left horizontal stabilizer.

The aircraft maintenance log indicated that on the previous flight the flight crew had written up a discrepancy that the number 2 engine would not go into reverse thrust. There was also a deferred maintenance item on the number 1 thrust reverser. In an interview with an FAA inspector, the company mechanic assigned to repair the number 2 thrust reverser (who reported his regular shift was from 4:30 A.M. until 1:00 P.M.) stated that when he finished work on the number 2 thrust reverser, he left the cowling wide open and asked the mechanic working on the number 1 thrust reverser to close his cowling when he was done and that he then signed off the number 2 thrust reverser in the maintenance log and left for the day. (This mechanic stated that he worked until 4:45 P.M. on this shift, 3 hours and 45 minutes past the end of his normal shift.)

Another company mechanic, who stated he reported for work at 3:00 P.M. on the day of the accident, told the FAA inspector that on the previous shift, he was scheduled to go off duty at 1:30 A.M. but actually worked until 8:00 A.M. This mechanic stated he then went home *but was unable to sleep* and reported back for his regular duty at 3:00 P.M. He stated that at about 3:30 A.M., he received a tie-in from the previous shift mechanic (assigned to the number 1 thrust reverser) that all cowlings on the DC-8 needed to be closed. He stated, however, that he did not review the tie-in log. This mechanic stated his primary duty that day was a Boeing 747 and that he did not get to the DC-8 until about 4:30 or 4:45 P.M. He stated that at that time, he noted that the cowlings for the number 1 and number 2 engines were closed and that those for the number 3 and 4 engines were wide open. He stated that he assisted in closing the number 3 engine cowl but did not check the number 1 or number 2 cowlings to ensure that they were secured. He subsequently reviewed the paperwork for the DC-8 and signed the airworthiness release for the aircraft. (A review of the aircraft logs disclosed no specific documentation that the number 1 or 2 cowlings had either been opened or closed.)

NTSB and FAA investigators examined all recovered cowl sections. No cowl sections were attached to each other by any latch mechanisms, and no evidence of distress to any latches, latching pins, or associated areas was observed [italics added].

Sleep is a complex phenomenon generated actively by the brain. Although you and I can only understand our own sleep in the manner suggested by Sri Ramana Maharishi, sleep researchers have provided us with a strong quantitative understanding of sleep in general. We learned in the 1950s that it is not a passive vegetative state. Since then, we have learned that many environmental and behavioral factors may conspire to prevent us from obtaining adequate sleep. These include heat, noise, light, physical overexertion, time of day, anxiety, caffeine, nicotine, over-the-counter and prescription medications, and many others. Extremely fatigued individuals have no trouble falling asleep. However, even their sleep may be interrupted easily by factors such as those listed above, as in the case of the aircraft mechanic who was unable to sleep during the day, between work shifts. It is difficult for the brain to generate sleep during the day.

Two parallel systems of nerve cells in the lower parts of the brain are responsible for generating two of the main components of sleep, deep sleep and dreaming sleep. As sleep begins, the deep-sleep system causes cells in the cerebral cortex to increase their simultaneous activities. The results are, first, a quieting of electrical activity in cerebrocortical cells and a synchronization of cellular activity (around 14 hertz, or cycles per second) that we observe with brain wave (EEG) monitors through electrodes attached to the scalp. Then, there begins a slower synchronization of cellular activity (around 2 hertz) that lends another name to deep sleep: slow-wave sleep (SWS).

During the major nocturnal sleep period and most naps, the first 20 minutes or so of sleep is spent in a transition from drowsiness through the faster brain-electrical synchrony to the slower synchrony of deep sleep. Deep sleep may continue for 20 or 30 minutes. Then, there is a transition back to the faster synchrony and, often, up to drowsiness. This whole process takes about 90 to 100 minutes and is called a sleep cycle. A good night's sleep is usually composed of about five sleep cycles, or about 8 hours (5 cycles × 95 minutes ÷ 60 minutes/hour).

Sleep Biology and Napping 61

The structure of a sleep cycle changes as the night progresses. The first cycles contain more deep sleep than the later cycles that occur toward morning. The deep sleep occurs during the middle part a cycle. The cycles that occur toward morning contain more dreaming sleep than the earlier cycles. The dreaming sleep occurs during the ends of cycles. We usually awake from dreaming sleep at the end of the last cycle of the night. Of the total sleep time that occurs at night, we expect to see about 20 percent spent in deep sleep and about 25 percent in dreaming sleep in healthy, younger people. Interestingly, the amount of deep sleep observed in both naps and the first cycles of the night strongly depend on the amount of prior wakefulness, whereas the occurrence of dreaming sleep depends more on the circadian clock: Dreaming sleep tends to occur mainly during the predawn hours.

This nocturnal sleep cycle structure is delicate and is disturbed easily by the factors listed at the beginning of this discussion and by aging. Quantitatively, these sleep disturbances usually take the form of

- Reduced amounts of deep sleep
- Reduced amounts of dreaming sleep
- Increased numbers of awakenings and time spent awake during the night
- Inability to get to sleep quickly (increased sleep latency)
- Inability to remain asleep after a couple of cycles (sleep maintenance insomnia)

When these things happen to you, you probably perceive and report that the quality of recovery you have experienced through the sleep process is less than you desire.

Why do we sleep? We do not yet fully understand the functions of sleep. Of course, there are some learned speculations about the functions of sleep. Nocturnal sleep, in general, was probably adaptive: Because primitive humans were poorly equipped for night work, especially in terms of vision, they were best served by quiescence during the night. Attempts to hunt or forage at night could lead easily to injury or death.

We have speculated that deep sleep, especially during the first two sleep cycles of the night, may represent a period in which wear and tear on body tissues is repaired, especially muscle tissue. The main reason to think this is that there is a large, predictable release of growth hormone from the brain into the bloodstream associated especially with the first two sleep cycles of the night. In the fully grown adult, growth hormone supports the rebuilding of tissues, especially muscle. A competing theory suggests that deep sleep is associated with energy conservation, much like the torpor associated with hibernation. However, the actual metabolic saving achieved by 8 hours of sleep over nonsleeping rest is quite small.

We have speculated that one function of dreaming sleep may be a periodic, artificial stimulation of major body functions that keep it reasonably well prepared for action at night while the deep sleep process is active. The reason for this speculation was that dreaming sleep is characterized by extensive stimulation of body systems. Activity in the brain during dreaming sleep leads to cortical activity (dreaming), rapid eye movements (REMs), middle ear movements (MEMs), stimulation of the sexual organs, and both stimulation and inhibition of the major muscle groups. (A failure of coordination in the simultaneous stimulation and inhibition of the major muscle groups can lead to twitching and even thrashing during dreaming sleep. One sees this often in sleeping dogs. The REMs were the first characteristics of dreaming sleep to be observed and studied, by Nathaniel Kleitman and his students, Eugene Aserinsky and William Dement, at the University of Chicago Medical School in the 1950s, who named this kind of sleep. These observations, by the way, were the main clue that indicated that sleep was not just a passive state.)

There also seems to be a relationship, across species, between maturity at birth and the amount of dreaming sleep generated by the adult. There is more dreaming sleep in animals born relatively immaturely. Humans fall in the middle of this distribution. This relationship may have something to do with energy expenditure tradeoffs for mother and offspring during and just after gestation. However, the main question on many peoples' minds seems to be, How much should I sleep? People in the United States *reported* sleeping about 7 hours a night on the average in a National Sleep Foundation (NSF) poll taken in the year 2000. About one-third surveyed tended to sleep 8 or more hours, and one-third tended to sleep 6.5 hours or fewer.

These numbers are remarkably consistent with a systematic study across decades of age and across the two sexes conducted more than 25 years ago by Drs. Robert L. Williams, Ismet Karacan, and Carolyn J. Hursch, and reported in their book, *EEG of Human Sleep*, back in 1974. Using brain waves as the standard, with 10 or 11 males and 10 or 11 females in each age decade, they found the following numbers for how long people sleep[15]:

Age decade, years	Women, hours	Men, hours
20–29	7.2 (6.4 to 7.9)	7.0 (6.5 to 7.5)
30–39	7.1 (6.0 to 8.1)	7.0 (6.3 to 7.8)
40–49	7.1 (6.3 to 7.8)	6.5 (4.9 to 8.0)
50–59	7.2 (6.0 to 8.4)	6.5 (4.8 to 8.1)

The numbers in parentheses are the upper and lower values we would expect to encompass about 95 percent of the population, based on the data that were collected. I think it's interesting that, in our 40s and 50s, men tend to sleep less on average than women and younger men, and some men try to get by on about 5 hours of sleep per night. Is this biology or business? The data don't allow us to determine whether sleep biology prevent the older men from generating as much sleep as the younger men or the women, their jobs required less sleep for success at work, or some other factor was at work.

If these numbers hold true for commercial aircrews, the implication for the cockpit is that we would see more sleepiness in men in their 40s and 50s than in women and younger men. However, my impression from published field studies of aircrew sleep and fatigue is that older, more experienced pilots tend to manage their sleep habits on layovers better than younger, less experienced pilots do. The older pilots often report as much or more sleep and similar fatigue levels as their younger colleagues. Conversely, I speculate that older private pilots may not manage their rest time as well as older commercial pilots. Those who enter aviation at an early age tend to be those who then enter the field of commercial aviation. Those who enter aviation at a later age tend to have succeeded well enough in life to afford the financial costs of learning to fly. The work ethics of these older but relatively inexperienced pilots may lead them to the risky belief that working all day and flying all night is an acceptable approach to aviation.

Recent research has suggested that we need about 8 hours of sleep per night. That is the present recommendation from the sleep research community. Comparing this number to the reported 7-hour average sleep time and its associated ranges suggests that fully two-thirds of adults in the United States regularly sleep less than the recommended 8 hours. For some of these people, this may not be a problem in terms of daytime sleepiness and job performance. For many others, it is a problem, as they noted in the NSF poll.

Do you get enough sleep between flights or trips? That's a good question. Do people in the United States get enough sleep, in general? Another good question, one that has been discussed widely in the news media in recent years and a question discussed later in this book.

Flying Schedules and Sleep

Flying schedules have substantial effects on work demand, effort, performance, and fatigue. One aspect of these effects is that the level of crew member alertness and performance is governed strongly by the amount and quality of rest acquired before and between periods of work. There are three major determinants of sleep tendency during a period of intended wakefulness: (1) circadian-circasemidian effects, (2) the amount of preceding sleep, and (3) the length of time since the last sleep period.

Concerning the first determinant, there are well-documented, strong sleep tendencies for humans from midnight until dawn and, less so, during the midafternoon hours. The low points of the two-peak pattern of human error tendency illustrate for us those times during the day when it is difficult to sleep. Generally, it is difficult for the brain to generate sleep during the dawn to noon period and the late afternoon to late-evening period.

Inadequate sleep harms the attention and performance of all flight crew members. In the following case, a recent design change that was familiar to a flight attendant was ignored due to sleepiness and inattention. Old habits replaced the vigilance that was needed to deal with something new.[16]

> The main passenger cabin door opened shortly after takeoff at approximately 800 feet AGL. The airplane returned to the airport, and the door separated from the airplane while it was on final approach. All indications are that the door was closed and locked prior to takeoff. The lack of damage indicates the door was unlocked/unlatched when it opened. The airplane had a newly designed handrail/door handle installed which reversed the direction of motion to lock and unlock the door. Fourteen airplanes in the fleet incorporated this design. The newly hired #1 flight attendant had operated this handle on 16 of her 60 flights since being hired. She had been *on duty about 14.5 hours on the day of the incident with only 5 hours of sleep the previous night due to her flight schedule* . . . [italics added].

Commercial aircrew sleep and flight scheduling are two sides of the same coin. If an aircrew work-rest schedule were based solely upon the adequate daily recovery of human resources, through sleep, the amount of sleep acquired by the aircrew would be somewhat independent of the CDP. In fact, most flying schedules attempt to allow the single, major, allotted sleep period for each crew member to be at the recommended 8 hours. However, many actual sleep periods are determined by factors other than flying schedules. These include administrative demands, travel time to and from the airport, mix-ups with hotel accommodations, acquiring meals, mix-ups on show times, and so forth.

Flight scheduling in general aviation presents a different picture. Flights may or may not be incorporated into the workday. For those flights that are part of the workday, your general sleep debt may be of some concern. This is especially true when flights occur during the midafternoon hours. Your chronic sleep debt will interact with the midafternoon slump to impair your flying performance dangerously. For those flights that occur after work hours, your chronic sleep debt will interact with acute fatigue from the workday to impair flying performance. If the flight continues into the predawn hours, both your chronic sleep debt and acute fatigue will interact with the predawn error peak to bring you down. In addition, your all-night flight actually deprives you of sleep you need. In such cases, it's best to plan to sleep first and fly later. Personally, I would not even consider making a general aviation flight between midnight and dawn.

Conversely, working nights, trying to sleep days, and then trying to fly during the day can prove to be a problem. Consider this 490-hour, private-pilot scenario from 4 P.M. one afternoon in October of 1997[17]:

> The pilot stated that he had landed to the south in light and variable winds in visual meteorological conditions. The airplane swerved to the

right; he overcorrected, and the airplane ground-looped to the right, collapsing the right main landing gear and striking the right wing, bending the rear spar. In a written statement, the pilot noted that he "realize[d] that I had landed a little hot and had not completely stalled. I had relaxed back pressure on the elevator and had not planted my tailwheel. I feel that fatigue and a lack of attention to the job at hand caused this accident." He additionally noted that "I had been working the graveyard shift for the past seven nights and I was very tired. I should have gotten a few hours of sleep before the flight."

The chronic fatigue problems associated with shiftwork and night work are beyond the scope of this book. Many books and pamphlets are available on minimizing those fatigue problems. Suffice it to say that most shift and night workers do not recover from acute fatigue during their rest periods and accumulate fatigue from day to day. I speculate that they are much more prone to aviation incidents and accidents than day workers.

Cockpit Napping

Caveat: FAA rules do not at present allow cockpit napping in commercial aviation. However, the following information applies also to naps taken at home before flights and during layovers. Maybe, someday, it will apply also to cockpit napping in commercial operations.

Research and development conducted by NASA for the FAA over the last 20 years[18] has led to a recommendation for strategic cockpit napping for overseas flights to minimize the effects of long flight durations and multiple time zone shifts on pilot fatigue. Similarly, the Air Force has adopted cockpit sleep and napping strategies for bomber crews who use multiple aerial refuelings to fly 36-hour missions from air bases in the continental United States to distant targets and then return, without landing enroute.

We use the term *strategic* to emphasize the importance of preflight planning, as opposed to *ad hoc* napping enroute. To accomplish strategic nap planning, you need to sit down with the flight plan and make a table that looks like this example:

Time at home	GMT (Zulu) time	Elapsed time	Local darkness	Critical tasks	Best nap times
1700	0100	00	No	Departure	
1800	0200	01	Twilight		
1900	0300	02	Yes		
Etc.					

To determine local darkness at any point in the flight, use your best estimate or a computer program that shows the world map and the day-night pattern across the world. For the right-hand column, the best times for naps are when there are no planned critical tasks and the time at home is about 2200 to 0600.

As you travel east or west across time zones, your body clock will drift slowly from your home time. To adjust your estimate of body clock time during a trip across time zones

- Subtract 1 hour from home time for each 24 hour period spent in time zones west of your home time zone or

- Add 1 hour to home time for each 36-hour period spent in time zones east of your home time zone.

Take the plan with you and keep it handy in the cockpit—don't rely on memory. Plan to nap in accord with the plan.

The Air Force actually has Windows-based software that creates a sleep and napping schedule like the one shown above. The software is called the Aircrew Mission Timeline Tool (AMTT). The scheduling inputs that are needed are the flight-planned waypoints in latitude and longitude, Greenwich mean time for each waypoint, and your altitude at the waypoint.

How long should a nap be? You should plan to awaken before you reach deep sleep or after the deep sleep of the first sleep cycle. Thus, you should plan for a napping period of up to about 30 minutes (this includes time to fall asleep) or a sleep period of about 1½ hours. The reason for avoiding awakening from deep sleep is a problem we call sleep inertia. Most likely, you have experienced more than your usual grogginess upon awakening from sleep. When that happened, you probably awakened from deep sleep.

Sleep inertia is somewhat paradoxical in that your performance is worse after awakening than it was before you went to sleep.* We expect to experience performance recovery by sleeping, and yet we do not, at least for an hour or so. The things that we know about sleep inertia include the following:

- Inertia is greater immediately following deep sleep than immediately following dreaming sleep.

---

*Most of this information is from the excellent review article by Ferrara and De Gennaro.[19]

- The deeper the deep sleep or the more deep sleep experienced, the greater the inertia and its duration.
- Inertia after naps is greatest for naps that occur in the circadian nadir in the predawn hours.
- Attention and accuracy are more impaired by sleep inertia than are simple, automatic motor responses.
- These effects may be noticeable for up to an hour or more.

Some things we don't understand about sleep inertia include:
- The differences in sleep inertia between people
- The effects of motivation on countering sleep inertia

How may you deal best with sleep inertia in the aviation community? First, plan to be awake for an appropriate amount of time before scheduled critical tasks, including preflight planning. Second, try to avoid long periods of wakefulness. These will lead to increased amounts of deep sleep during the major sleep period and subsequent strong, long-duration sleep inertia. Naps between major sleep periods can help with this problem. If you are not sleep deprived or only mildly sleep deprived, your sleep inertia will last 5 to 30 minutes. If you plan to be awake for at least 30 minutes before the task, you should not have a problem performing the task. If you have missed all or most of a night of sleep, your sleep inertia could last ½ to 2 hours after a nap of up to 45 minutes, or even 3 hours after a nap of 90 minutes. Plan accordingly. Finally, if you are sleep deprived, you may wish to avoid napping at all during the predawn hours if you must complete a critical task within a couple of hours after the nap. You will find it very difficult to recover from sleep inertia in that situation.

5

Sleep in the News

Much has been written in magazines and newspapers lately about various diets as sleep aids. In part, the newsworthiness of this topic has been driven by the perception that we have a national sleep debt in the United States. That is, few of us sleep as much as we need to. I have examined here some current data about how much sleep we get in the United States and speculated about aircrew sleep, in comparison. Then, I have consolidated a large amount of information about the effects of nutrition on sleep and presented a summary view.

Leading the charge to inform the public of our general sleep debt is the nonprofit National Sleep Foundation of Washington, D.C. (NSF; *www.sleepfoundation.org*). Here are some relevant findings from their most recent sleep poll, excerpted from the executive summary of their report about the poll:

> A national survey of American adults, the National Sleep Foundation's 2000 Omnibus Sleep in America Poll (OSAP) examined the public's

beliefs and habits regarding sleep, and the consequences of these beliefs and habits. The 2000 poll focused on the effects of sleep habits and attitudes in the workplace, particularly on performance. It also examined . . . drowsy driving and the prevalence of sleep problems and disorders among American adults. The OSAP is a nationally representative telephone survey of 1,154 adults living in households in the continental U.S. With 95 percent confidence, estimates based on this sample size have an error range of plus or minus 3 percentage points.

Most adults in the U.S. get less sleep than they need based on sleep experts' recommendation of 8 hours per night.

- On average, adults sleep 6 hours and 54 minutes during the workweek, about an hour less than the 8 hours recommended by sleep experts. Most adults compensate for their sleep loss during the workweek by sleeping longer on the weekend, with an average increase of about 40 minutes.
- Only one-third (33 percent) of adults say they get at least the recommended 8 hours or more of sleep per night during the workweek; and
- One-third (33 percent) of adults say they get fewer than 6.5 hours of sleep per night during the workweek.

Almost one-half of all adults (45 percent) agree strongly or somewhat with the statement that they will "sleep less to get more work done." [Historically, this statement has been a tenet in military and commercial aviation.]

More than one out of ten adults (13 percent) report that sleep is the first thing they give up as compared to time with family/friends, recreational activities, and household/personal chores when they do not have enough time to get everything done. [This is the "Let's party now, I can sleep when I'm dead" attitude. Unfortunately, in aviation, this particular effect of not sleeping may occur sooner rather than later.]

More than four out of ten adults (43 percent) agree that they "often stay up later than they should because they are watching TV or are on the Internet."

Slightly more than one-half (52 percent) of adults agree that they "need an alarm clock to get up in the morning." [For people with regular day jobs, your body clock should wake you at the same time each morning. However, for aircrews with irregular duty hours, alarm clocks are usually a necessity.]

Nearly two-thirds of adults in the U.S. (62 percent) experienced a sleep problem a few nights per week or more during the past year.

- Specifically, more than one-half of the adults surveyed (58 percent) report having experienced one or more symptoms of insomnia a few nights per week or more within the past year. [Probably much higher for commercial aircrews.]

- The most common symptom of insomnia experienced a few nights a week or more was "woke up feeling un-refreshed" (43 percent), followed by: "were awake a lot during the night" (34 percent), "had difficulty falling asleep" (22 percent), and "woke up too early and could not go back to sleep" (22 percent). [These are common complaints among commercial aircrews who must sleep during the day.]

- "Waking up un-refreshed" is more likely to be reported by those . . . working more than 60 hours per week. [As do most successful people in business and many commercial aircrews.]

At the top of the list of factors disrupting sleep a few nights per week or more was "Stress," reported by 22 percent of adults overall, 26 percent of women and 16 percent of men. In addition, there were "Environmental factors (noise, light, or temperature)," reported by 16 percent of adults overall; "Bedding," reported by 14 percent of adults overall; "Nasal congestion," reported by 12 percent of adults overall; "Allergies," reported by 11 percent of adults overall; and "Indigestion," reported by 8 percent of adults overall. [There were also sleep pathologies and pathology indicators reported, but I have not listed them here. The items I have listed are all highly likely to be experienced by general aviation pilots and aircrews.]

A sizable proportion of adults (43 percent) report that they are so sleepy during the day that it interferes with their daily activities a few days per month or more; and, one out of five (20 percent) experience this level of daytime sleepiness at least a few days per week or more. [Are you in these groups? You can estimate your daytime sleepiness, below.]

These numbers certainly support the idea that we do, indeed, have a national sleep debt. In addition, most of these responses presumably came from people with regularly scheduled daytime jobs (I have omitted the poll responses from shift workers). Aircrews usually work very irregular hours and schedules, making them even more susceptible to the problems listed.

How sleepy are you? To answer this question, the National Sleep Foundation asked responders to rate their daytime sleepiness on the Epworth Sleepiness Scale (ESS). I have used it myself in various research studies, and it is quite useful. M. W. Johns of the Epworth Hospital in Melbourne Australia published the ESS in the open research literature in 1991 and 1992. It was designed to help sleep clinicians decide whether a person has a sleep pathology such as sleep apnea or narcolepsy. In the clinic, scores greater than 15 on this 0 to 24-point scale are taken to indicate that the person does have a sleep pathology, and scores of 10 or less to suggest no pathology. For the poll, the NSF included scores in the gray, 10- to 15-point area as indicating too much daytime sleepiness. They reported that

> Further evidence for the high level of daytime sleepiness experienced by U.S. adults is found in the one-third (32 percent) who suffer from significant daytime sleepiness as measured by a score of 10 or more on the Epworth Sleepiness Scale (ESS).
>
> Not surprisingly, more than one-third (37 percent) of those adults suffering from daytime sleepiness say they are not satisfied with the amount of sleep they get during the workweek, compared to one out of five (18 percent) adults without significant daytime sleepiness.

If you would like to estimate your daytime sleepiness, here is the Epworth Scale, as presented by the NSF. Add your ratings of zero through three across the eight situations to get your daytime-sleepiness score.

The following questionnaire will help you measure your general level of daytime sleepiness. Answers are rated on a reliable scale called the Epworth Sleepiness Scale (ESS)—the same assessment tool used by sleep experts worldwide.

Each item describes a routine daytime situation. Use the scale below to rate the likelihood that you would doze off or fall asleep (in contrast to just feeling tired) during that activity. If you haven't done some of these things recently, consider how you think they would affect you.

Please note that this scale should not be used to make your own diagnosis. It is intended as a tool to help you identify your own level of daytime sleepiness, which can be a symptom of a sleep disorder.

Use the following scale to choose the most appropriate number for each situation:

0. Would *never* doze
1. *Slight* chance of dozing
2. *Moderate* chance of dozing
3. *High* chance of dozing

Situations:

Sitting and reading	_____
Watching TV	_____
Sitting inactive in a public place; for example, a theater or meeting	_____
As a passenger in a car for an hour without a break	_____

Lying down to rest in the afternoon
when circumstances permit   _____

Sitting and talking to someone   _____

Sitting quietly after lunch without alcohol   _____

In a car while stopped for a few minutes
in traffic   _____

The poll also disclosed that

> Of U.S. adults, 6 percent have used medications to stay awake. Of those who use a medication to stay awake, the vast majority (85 percent, or 5 percent overall) use over-the-counter medications, while about one out of ten (12 percent, or 1 percent overall) use prescription medications and almost no one uses both.

I discuss pharmacological countermeasures for fatigue and sleepiness later in this book. Do you fly when you are sleepy? If the poll's data on drowsy driving can be extrapolated to flying, I'm sure you do:

> About one-half of adults in the U.S. (51 percent) report driving while drowsy in the past year; nearly one out of five (17 percent) have actually dozed off while driving. [Ever doze off in the seat?]
>
> About four out of ten adults (42 percent) say they become stressed while driving when drowsy.
>
> The majority of adults (63 percent) use caffeine when they feel drowsy while driving.
>
> Approximately one out of five drivers (22 percent) report that they pull over to nap when they feel drowsy while driving. [More from the poll about napping at work, below.]

How do work and sleep interact? The poll indicated the following:

Only one out of ten adults (10 percent) are not satisfied with the amount of sleep their job allows them. [I suspect that this proportion is quite a bit higher among commercial aircrews.]

One-half of adults (51 percent) say they are "completely satisfied" with the hours of sleep their work schedule allows. They say they need an average of 7 hours and 6 minutes of sleep not to be sleepy at work and are actually getting an average of 6 hours and 53 minutes of sleep during the workweek. However, this is *still one hour less than what experts recommend for optimum well-being.* [Emphasis theirs. Data I have collected tend to agree, suggesting that 7¼ to 7½ hours is the desired amount of sleep during the night to be sharp on the job during the day.]

Women say that they need an average of approximately a half an hour more sleep than men to not be sleepy at work (7 hours 18 minutes and 6 hours 48 minutes, respectively).

Despite their perception that they are getting approximately the amount of sleep they feel they need, *daytime sleepiness interferes with many adults' work* [emphasis theirs].

- More than one-fourth of adults (27 percent) say that they are sleepy at work 2 days per week or more. Those reporting higher rates of sleepiness at work include . . . those working more than 60 hours per week.

Do you relate to any of the following statements about sleepiness at work?

Nearly one out of ten adults (8 percent) say that they occasionally or frequently fall asleep at work.

One-fourth of adults (24 percent) say that they have difficulty getting up 2 or more workdays per week. [This is certainly true for commercial aircrews who must report for work in the midnight to dawn period.]

Some adults (4 percent) say they have not gone to work due to sleepiness occasionally or frequently. [I wonder if this is about the same frequency with which aircrews declare the need for additional crew rest and general aviation pilots decide not to fly.]

More than one out of ten adults (14 percent) say that they are occasionally or frequently late to work due to sleepiness.

One out of five adults (19 percent) report that they occasionally or frequently make errors at work due to sleepiness, especially . . . adults that work more than 60 hours per week. [I wonder if this less than the frequency for minor aircrew errors.]

Two percent of adults report that they occasionally or frequently are injured at work due to sleepiness.

Nearly one-half of those surveyed (46 percent) say that working more than 10 hours in a day makes them too sleepy to do quality or safe work. This finding is cut in half for working 8 hours or less (25 percent) or 8–10 hours (24 percent). [How many of commercial aviation trips are less than 10 hours long?]

In a separate question, respondents were asked whether sleepiness interferes with their ability to function in their jobs. For respondents who stated that their work was diminished

- More than six out of ten (61 percent) say their concentration is diminished. [Research indicates that the ability to sustain attention is one of the first things affected by fatigue and sleepiness.]
- More than one-half (51 percent) say sleepiness interferes with the amount of work they do.
- Four out of ten (40 percent) say the quality of work they do suffers.
- Of those who said that sleepiness interferes with the ability to function in their job, on average, workers report that their ability to concentrate, the amount of work they do, and the quality of their work are each *diminished* by about 30 percent when they are sleepy at work. [My rule of thumb for pilots is that a reduction beyond *10 percent* may be unacceptable.]

In addition, how does work, in turn, affect sleep?

About one out of ten adults (8 percent) report that *work problems disturb their sleep* 2 or more days per week or more. [Emphasis theirs. For commercial aircrews, jet lag and night work would fit in here.]

Almost one out of ten adults (7 percent) report having *changed jobs in order to get more sleep.* [Emphasis theirs. Commercial aviation, like shiftwork, appears to cause self-selection. Those who deal poorly with the fatigue caused by jet lag and night work tend to leave the profession very early after starting work.]

I discussed cockpit napping in the preceding chapter. How did these respondents feel about napping at work?

About one out of six adults (16 percent) report that their employers allow them to take naps at work.

- Nearly one-half (46 percent) of those who are allowed to nap at work do so.
- One-third of adults (33 percent) say that they would nap at work if it were allowed.
- One out of ten adults (10 percent) nap before going to work.
- More than one-third of adults (35 percent) nap after work.

What do people do about sleepiness?

When experiencing difficulty sleeping, about one-half of adults (53 percent) report that their "most likely" reaction is to do nothing about it. [Historically, this has been the practice in aviation.]

Almost one out of five of adults are most likely to use herbal remedies (18 percent) or over-the-counter medications (17 percent), while 7 percent are most likely to use prescription medication when experiencing difficulty sleeping.

Only one-fourth of adults (24 percent) agree that "if they had trouble sleeping for a month, they would take a sleeping pill."

Hesitancy to take sleeping pills is explained, at least in part, by the finding that four out of ten adults (40 percent) agree that "if they start using sleeping pills, they might always need them to sleep."

When specifically asked about their use of medication to help them sleep in the past year, nearly one-fourth of adults (22 percent) say they have taken medication to help them sleep in this time period. Of those who have used a medication to help them sleep:

- Most used over-the-counter medications (56 percent);
- Some used prescription medications (35 percent); and
- A few used both (10 percent).

One out of five adults in the U.S. (19 percent) have used alcohol to help them sleep. [In the past, this proportion has probably been much higher among some aircrews.]

As I mentioned earlier, I discuss some pharmacological options for dealing with sleep problems in a later chapter. There, you will learn that the alcohol option is not a good one if you are seeking good sleep quality.

This poll shows us that, in general, we have a strong national tendency to not sleep enough. When aircrews are faced with irregular schedules that include a lot of night work and jet lag, they face a three-pronged fork in the road. You may take one of three paths. First, you may do nothing about the impending sleep disruption, as would 53 percent of those polled. Second, you may undertake behaviors that make you more fatigued ("I can sleep when I'm dead"), and third, you may undertake behaviors that make you less fatigued. I have attempted in this book to present you with many things you can do to make yourself less fatigued. One of those is to eat well.

Nutrition

There has been a lot of information published lately in books and newspapers about what to eat to help you sleep. Here is my take on the literature.

It is possible that complex carbohydrates are the most suitable food to have as your last meal before your major sleep period. A meal high in complex carbohydrates—polysaccharides such as grains, breads, pastas, vegetables, fruits—may act to increase the circulating levels of an amino acid called L-tryptophan. L-tryptophan is, among other things, a metabolic building block of a sleep-related, cell-to-cell transmitter in the brain. Sleepiness is common after high-carbohydrate meals.

Warm milk is a classic bedtime drink. It is supposed that this is true because milk contains L-tryptophan. This may be the case, although the effect has been difficult to detect in the laboratory. However, lactose, a disaccharide found in milk, must be broken down by lactase in the small intestine. Lactase deficiency (low intestinal levels of the lactase enzyme) leads to diarrhea, bloating, and flatulence after the ingestion of lactose (lactose intolerance). These sleep-disrupting symptoms are reasonably common, especially in those of Asian descent, and may increase with aging. The use of lactose-free dairy products (such as McNeils' Lactaid milk) or lactase supplements in pill form (such as Lactaid pills or Rite-Aid's Dairy Ease) can be helpful. There is a lot of information on the Internet about these products; this is a common problem.

We usually take in sugar in candy and other sweets as the monosaccharides, glucose and fructose. These sugars act as mild stimulants and may affect sleep patterns. Avoid them before your major sleep period.

Many of us do not eat enough fiber. We need about 20 to 30 grams of fiber per day. Adequate dietary fiber may help minimize the potential for digestive disorders and cancers of the digestive tract. However, although high-fiber foods such as beans, raw onions, cabbage, and cauliflower are all good for you, they may leave you bloated, making sleeping difficult. If you don't eat enough fiber now, add it in small amounts to your diet so your body can adjust to it gradually, creating more of the enzymes needed in the intestines to break down the fibers. Avoid eating relatively large amounts of high-fiber and gas-producing foods before your major sleep period. If you must eat them, products containing galactosidase (such as Block's Beano) taken with the food, can help support their digestion. After the meal, antigas products containing simethicone (such as Novartis' Gas-X) can help offset these problems, allowing you to sleep.

Fats are difficult to digest. A high-fat meal may remain in the stomach for 2 hours or more, generating some physical discomfort. For those prone to heartburn, a burning sensation in the esophagus, near the heart, or gastric reflux disorders, in which stomach acid attacks the mucous lining of the esophagus above the stomach, large or high-fat meals can be a real problem. Similarly, acidic foods, containing tomato products or hot spices or eating too fast may cause heartburn. Lying down may make heartburn worse. The discomfort of heartburn can prevent sleep onset and wake you up in the middle of the night. An acid controller (such as SmithKline Beecham's Tums or Merck's Pepcid AC) can help combat this problem and let you sleep.

On the other hand, a piece of advice about fats caught my eye recently and may be helpful to some. The endocrinologist, Marshall Goldberg, interviewed by Marilyn Marter of the Knight-Ridder News Service, pointed out the usefulness of small amounts of fats in reducing feelings of hunger. Fatty acids in the duodenum stimulate the release of cholecystikinin (CCK) from cells lining the duodenum. The main function of CCK is to stimulate pancreatic secretion. However, it also acts through receptors in the gut that relay signals to the brain to inhibit food intake. Hunger pangs are reduced in this process. Goldberg pointed out that you might use small amounts of the good fats—monosaturated fats—to accomplish this. These good fats are found in large proportions in olive, almond, and hazelnut oils and, to a lesser degree, in canola, peanut, and sesame oils. He has had good success recommending olive oil as an aid in weight control. I have found the oil quite rich when taken straight. However, it appears that you should consider a moderately sized pasta meal made with olive oil to be a healthy and sleep-friendly component of your diet.

Eating just a big meal may make you feel drowsy, but the time and effort it takes to digest it may keep you awake. However, don't starve yourself before you try to sleep either. When blood glucose falls to a low level, the pancreas releases a hormone called glucagon, which sends a signal to the liver, to move more sugar into the bloodstream (the opposite effect of insulin). However, glucagon also has cardiovascular effects, similar to those of adrenaline, the fight-or-flight hormone: Heart rate and blood pressure go up. This is not conducive to restful sleep.

Conversely, starving yourself during the workday can, in some situations, lead to low blood sugar levels and poor cognitive performance. Consider this hapless, 3400-hour instructor pilot's experience at about 5:30 P.M. in August of 1999[20]:

> The airplane landed wheels up after the instructor pilot failed to lower the landing gear. The instructor told the student to execute "a no flap landing due to a simulated hydraulic pump failure." The student established the airplane on the approach and called for the "emergency gear extension checklist." The instructor delayed extending the gear in accordance with the operator's flight standards manual, which stated that the landing gear should not be extended until the landing was assured. Later in

the approach, when the gear warning horn stopped sounding, due to the student's movement of the power levers forward, the instructor removed his hand from the gear handle without extending the gear. The instructor stated that "because [the student] had already called for the [emergency gear extension] checklist once before, in a split second thought process, [he] mistakenly thought it had been completed." Following the accident, the landing gear system was tested and found to operate normally. Review of the maintenance records revealed no uncorrected discrepancies. At the time of the accident, the instructor pilot was completing a 9-hour work day, and did not have a lunch break.

The NTSB concurred about the probable cause[20]:

> The instructor pilot's failure to complete the emergency gear extension checklist, resulting in the inadvertent wheels-up landing. A factor was the instructor pilot's fatigued condition.

Here are a few hints for stocking the larder at home for those who do not have their own or another's cooking skills:

- Make a meal plan that emphasizes proper nutrition. Then, according to the plan, keep your cupboard stocked with healthful foods instead of fast foods and fatty or sugary snacks.
- Don't leave home with an empty cupboard to greet you on your return. The empty cupboard only encourages you to turn to fast food.
- Stock healthy, portable snacks; for example, small boxes of raisins, that you can use at home

to avoid low blood glucose levels. Also, carry these on trips. Breakfast or protein bars are useful, also.
- Supplement frozen meals with fresh fruits and fresh or fresh frozen vegetables. Frozen meals may be lacking in vitamins and fiber. Salads are easy to make, and the supermarket even offers presliced makings.

The bottom line for nutrition is to
- Maintain good nutrition—adequate levels of protein, complex carbohydrates, and dietary fiber and small amounts of monosaturated fats
- Watch for symptoms of lactose intolerance
- Before sleep, avoid
  - Large meals
  - High-fat meals
  - High-acid meals
  - Sweets
  - Hunger
- If you eat before sleep, emphasize grains, breads, pastas, vegetables, and/or fruits

One last note: Although all agree that adequate hydration is essential for well being, we researchers have not yet systematically examined the effects of dehydration on sleep, itself. The results of one study I've found suggested that dehydration might interfere with the production of deep sleep. At the very least, drink plenty of water every day—about 1 ounce of water for every 2 pounds of body weight, or about 30 milliliters per kilogram of body weight. That's 50 ounces of water, or about 1.5 quarts, per hundred pounds of weight (about 1.5 liters per 50 kg of body weight).

This advice is even more important for commercial aircrews than for the office workers to whom it is directed. Because aircraft fly in relatively dry air at altitude, from which aircraft air conditioning systems remove even more water, commercial aircrews work in extremely dry environments while in flight. I'd guess that aircrew who spend a lot of time in flight at altitude have dehydration problems similar to those of serious athletes who sweat a lot during workouts. Heeding hydration advice provided for athletes would probably not be a bad idea for aircrews, at least in terms of the volume of fluid you need to replace. The one difference is that, with exercise and sweating, glucose is burned and sodium is lost. Athletes replace the glucose and sodium with isotonic drinks, the prototype of which is Gatorade. Aircrews just need plain water.

6

Dealing with Jet Lag

The following report was captured by NASA's anonymous Aviation Safety Reporting System:

> After seven days in the Pacific, we fly all night from Bangkok to Narita, have a short day layover, then fly all night to Honolulu. Some or all of the crew passes out on the last leg from fatigue. We are so tired by the approach and landing that our thinking and reaction times are similar to being drunk. If the weather wasn't consistently good in Honolulu, I'm sure we would have lost an airplane a long time ago.

I have described how circadian and circasemidian rhythms combine to produce the predawn and midafternoon peaks in human error probability and occurrence. Here, I have described how these rhythms are generated in the body and how they, mainly circadian rhythms, interact with time zone changes and night work to produce the malaise we experience as jet lag.

The impact of a human circadian rhythm that is not aligned with the local daylight-darkness cycle is familiar to anyone who has suffered jet lag. Consider the inability of an individual from the West Coast of the United States to awaken refreshed on the first morning of a sojourn on the East Coast. At 6 A.M. on the East Coast, the brain's circadian rhythm is operating as if it were 3 A.M. It is trying to generate sleep activity, reaction times are slowed, aerobic physical capacity is slightly impaired, the expected frequency of job errors may be many times the expected frequency at noon, and so forth. The person suffers a general perception of malaise. An individual from the United States visiting Europe will have similar problems. We suffer the general malaise associated with jet lag until the body's circadian rhythms realign with the local daylight-darkness cycle.

To explain these problems, I start with a brain-geography mini-lesson. You may know that some of the optic nerves cross when they leave the retinas at the backs of the eyes, on their way to the left and right sides of the optic cortex at the back of your head. The crossing, or chiasm, occurs above the roof of the mouth. Just above this crossing, in the bottom of the brain, lies the thalamus. The thalamus is the upper end of the regulatory systems that deal with automatic functions in the body. Many hormones are released and regulated from this area, especially its lower part, the hypothalamus. Within the hypothalamus, and just above the optic chiasm, lies a small group of nerve cells called the suprachiasmatic nucleus, or SCN. The SCN seems to be the main regulator of metabolic circadian rhythms in the body.

There are many circadian rhythms in the body. They include daily fluctuations in body temperature, metabolic rate, various hormones, dreaming sleep, strength, aerobic power, and most kinds of cognitive function. It seems likely that the circadian rhythm in metabolic rate, manifested reasonably well as body temperature, underlies the rhythms in physical and cognitive functions.

We have no firm idea why these cycles exist. All of the cycles are linked strongly to the daylight-darkness cycle for those who work days and sleep nights. This linkage seems to be provided mainly by daylight and darkness information transmitted to the SCN, and to a much lesser degree by social cues. The main social cues probably include meals and physical activity associated with work and leisure.

There is also a pacemaker in the SCN. It generates an approximate circadian rhythm that is a little more than 24 hours long. Much like many pacemaker cells in heart muscle, the pacemaker cells in the SCN seem to be influenced easily by faster rhythms. Thus, apparently, when the slightly faster 24-hour daylight-darkness cycle is sensed, the pacemaker in the SCN slaves itself, or becomes *entrained*, to the faster rhythm. Conversely, when all external time cues, especially daylight-darkness and social cues, are removed or randomized, the SCN pacemaker cells often take up their normal rhythm of slightly more than 24 hours.

When all of the cycles are linked to the daylight-darkness cycle, they parallel each other in their daily pattern of highs and lows. Body temperature, the stress hormone cortisol, and physical and most cognitive functions reach very low troughs in the predawn hours (which, presumably, is why we are at quite high risk for making errors during the predawn hours). They rise through the morning hours, and then the rise stops and flattens out in the midafternoon. The flattening is caused by the rhythmic interaction, or modulation, of the circadian rhythm by the circasemidian rhythm. Workers often refer to the flat spot as the "postlunch dip." This midafternoon slump in performance and alertness has little to do with lunch, although a large, high-fat lunch can make it worse. The midafternoon flat spot is caused mainly by normal body rhythms.

Body temperature, cortisol, and physical and cognitive functions then increase for a short time before reaching a peak somewhere in the period, 6 to 10 P.M. The time of occurrence of the peak, called the *acrophase* of the circadian rhythm, depends mainly upon the part of the night (early, late) the individual sleeps. After reaching acrophase, temperature, cortisol, and physical and cognitive functions start down toward the predawn trough. The predawn trough usually occurs about 9 hours after the acrophase.

For those who remember their trigonometry from high school or college, the pattern I have described can be modeled quite well. You use a 24-hour period sine wave with its peak at about 6 P.M. and a smaller amplitude of a 12-hour period sine wave with peaks at about 10 P.M. and 10 A.M. In fact, we use this method to model circadian and circasemidian rhythms. In addition, of course, we add recovery (in sleep) and wakefulness functions to those models.

You now have a picture of the expected circadian rhythm for a person who works days and sleeps nights: The high point occurs in the late afternoon or early evening, the low point occurs during the predawn hours, and there is a secondary low point during the midafternoon. There are differences between people and even within individuals due to varying times of day work and night sleep. For example, the acrophase of a person who starts work at 6 A.M. may occur about 3 hours earlier than the acrophase of a person who starts work at 9 A.M. In addition, if the person who starts work at 6 A.M. sleeps in on the weekends, he or she may experience a 1- or

2-hour drift in acrophase to a later time. This sets up a jet lag effect on Monday morning.

Why would the person who sleeps in on weekends experience a jet lag effect? More generally, why do we experience this malaise when we move quickly, by air, from one time zone to another? The answer seems to lie in the observations that (1) the SCN pacemaker is entrained to the local daylight-darkness cycle and (2) the SCN pacemaker's natural rhythm is slightly longer than 24 hours.

Let's consider the late Monday sleeper first. Assume that this person has to be at work at 6 A.M. Monday through Friday and that his or her circadian acrophase occurs at about 6 P.M. and the trough occurs at about 3 A.M. Assume also that he or she goes to bed several hours later than usual on Friday and Saturday nights and sleeps in several hours on Saturday and Sunday mornings. Although the local daylight-darkness cues have not changed, the timing of sleep and activity has shifted later on the clock. This shift confuses the SCN pacemaker slightly, and it may tend to free-run at a slightly longer period than 24 hours. Thus, the person's acrophase may drift 1 or 2 hours to 7 or 8 P.M. by Sunday. Now, the predawn trough has drifted from before work to occur during work. Experiencing the trough in body temperature and physical and many cognitive functions while we are awake, instead of sleeping through the trough, seems to embody the malaise we report as jet lag. You can see that sleeping in on days off may not be a good thing for those who usually arise early for work. What's their alternative? Get up at the same time every day, even on days off. Then, if more sleep is needed, take a midafternoon nap.

Now, let's consider jet lag as an extension of the Monday morning problem. In scheduled, commercial aviation, we are able to change time zones at the rate of about one zone per hour. It is common to make sudden changes (that is, sudden for the SCN clock) of 3 hours and more. What happens to circadian rhythms when we make such a change?

If we travel three time zones east in 3 hours, we will not only shift the timing of sleep and activity earlier on our SCN clock, the SCN clock will also experience earlier local daylight-darkness cues. For example, let's say that we don't sleep in on the weekend. Instead, for whatever reasons, we go to bed at our weeknight times on Friday and Saturday nights and get up at our weekday times on Saturday and Sunday mornings. On Sunday, we fly from the West Coast to the East Coast. When it's bedtime on the East Coast, our West Coast SCN clock thinks it's still daytime, or early evening. We probably can't get to sleep. Then, when it's time for us to get out of bed to be at work at 6 A.M. on the East Coast, our West Coast SCN clock hasn't even hit its trough yet. Thus, similar to the late sleeper, we experience the trough in body temperature and physical and most cognitive functions while we are awake, instead of sleeping through the trough, and we perceive the malaise of jet lag. Conversely, if we travel three time zones west in 3 hours, when our East Coast SCN clock is ready for bed, it's still daytime or early evening on the West Coast. We may or may not sleep. Then, when it's time for us to get out of bed to be at work at 6 A.M. on the East Coast, our East Coast SCN clock wakes us up. However, it's the middle of the night on the West Coast.

These are the acute, or short-term, effects of rapid time-zone changes. What happens in the longer term, over a period of several days or weeks after making the time zone change? In those situations, we must contend with the metabolic trough that occurs during waking instead of occurring during sleeping, and we must try to sleep at flight-schedule-dependent times that have nothing to do with the signals coming from the SCN or the times of local daylight and darkness.

The sleep and circadian rhythm researcher, Dr. Martin Moore-Ede of Harvard, describes in his book, *The Twenty-Four Hour Society*, an example of how jet lag may trigger an accident. Here is a summary of the schedule flown by the accident aircraft, a Pan Am Boeing 707 that crashed in Bali in 1974, killing all on board[21]:

Time	Activity
7:44 P.M.	Depart San Francisco
1:32 A.M.	Arrive Honolulu
3:39 A.M. the next morning	Depart Honolulu
2:35 P.M.	Arrive Sydney
6:21 P.M. the next day	Depart Sydney
1:30 A.M.	Arrive Jakarta
2:18 A.M. the same night	Depart Jakarta
6:40 A.M.	Arrive Hong Kong
4:00 A.M. the next morning	Depart Hong Kong
8:30 A.M.	Crash at Bali

Although the layovers at Honolulu and Jakarta were long, they emphasized sleeping during local daylight hours and with offsets from the SCN clock. I wonder how effective that sleep was? Considering the result, I can guess that the sleep at those two locations did not provide full recovery for the aircrew.

Dr. Moore-Ede makes the point that such schedules seldom result in crashes. Most of the time, aircrews have flown such schedules without a reportable accident. However, when the joint probabilities of fatigue-reduced alertness and attention collide with the need to detect and avoid flight hazards, the aircraft is likely to collide with the terrain. We can't move the terrain, but we can try to minimize aircrew fatigue.

Before contemplating the minimization of the effects of continual time zone changes on aircrew sleep, fatigue, and performance, you need to understand a bit about how you may help the body respond to single-step time zone changes. This is the kind of advice given to the business traveler.

Can you acclimatize to the new time zone?* The answer is Yes; you will acclimatize naturally, but slowly. In the meantime, you experience the symptoms of jet lag, although they decrease each day during acclimatization. However, there may be things you can do voluntarily to speed up the acclimatization process.

Two recently recommended methods for speeding up one's acclimatization to a new time zone include the uses of melatonin and bright light, together or independently. Briefly, melatonin, a hormone produced by the brain, helps induce sleep. Its concentration in the blood is much higher during sleep at night than at other times of day. Melatonin supplements are advertised widely as sleep aids. Bright light tends to emulate the effects of daylight, suppressing melatonin secretion in the brain and reducing sleepiness.

Caveat: The jury is still out on the effectiveness of melatonin ingestion for these purposes. Some studies suggest that early reports of supplemental melatonin effectiveness for dealing with jet lag may have been due to placebo effects. Another study suggests that, instead of accelerating the body's change to a new time zone, melatonin supplementation may only help stabilize the body's change as it occurs. In addition, questions about

---

Acclimatization is the word physiologists use to describe semipermanent physiological adjustments to a new environment. Writers tend to use the word *adaptation*. However, for physiologists, adaptation refers to genetic change. Because we are talking about physiological changes in the SCN, I opt here for *acclimatization*.

the long-term safety of melatonin supplements have not yet been answered. Melatonin is, after all, a naturally produced hormone that may have small effects on many body functions besides sleepiness.

The following recommendations for using melatonin to combat jet lag are based, in part, on recommendations from the chronobiology researcher, Dr. Alfred Lewy. One fundamental aspect of the technique he suggests is to use a very small dose of melatonin to induce earlier or later sleepiness just before and just after time zone changes. If there are unknown, long-term deleterious health effects associated with taking melatonin, this use of a very small dose should help ameliorate them.

Consider first our West Coast resident who travels to the East Coast. The traveler would take 0.5 milligrams of melatonin at his or her intended East Coast bedtime on the several days leading up to the trip. Evidence that this is working would be that you wake up somewhat earlier than usual on the mornings just before the trip. The traveler from the United States to Europe could use this same technique. The time that he or she takes the melatonin at home would be earlier because of the greater time difference. In addition, the traveler could take 1 to 3 milligrams before bedtime the first night in the new time zone to help induce sleep.

Now consider our East Coast resident who travels to the West Coast. The traveler would take 0.5 milligrams of melatonin at his or her West Coast at home time of awakening on the several days leading up to the trip. This practice should encourage later sleeping. After time zone changes of more than 3 hours westward, the traveler would also take 0.5 milligrams upon awakening in the middle of the night or in the morning for several days in the new time zone. In addition, the traveler could take 1 to 3 milligrams before bedtime the first night in the new time zone to help induce sleep. Most people working with melatonin suggest that, because it may make you sleepy, you should experiment with it safely at home before taking it on the road. Try several doses from 0.5 to 3 milligrams at home on days when you know you do not need to be alert.

Dr. Terry Lyons of the USAF Office of Scientific Research nicely explained the philosophy behind the use of bright light for dealing with jet lag:

> Exposure to light is the strongest driver of the internal body clock. For westward travel, light exposure before the middle of the night (departure location) temperature minimum/melatonin peak (in most people occurs between midnight and 4:00 A.M.) should help in achieving a Phase Delay of the circadian pacemaker. For example, at 2:00 P.M. Tokyo time it is already midnight Washington, D.C. time. Sunlight exposure starting before 2:00 P.M. Tokyo time would be very advisable after arrival in Tokyo—bright (outdoor) light is best. This should help to shift/delay the temperature minimum/melatonin peak to align it better with actual Tokyo nighttime. For eastward flights, avoid bright lights upon arrival at your destination until after the temperature minimum/melatonin peak has passed. For example, when leaving Tokyo travelling to Chicago it could be expected that the temperature minimum/melatonin peak entrained to Tokyo time would naturally occur at around noon Chicago time. Morning bright light exposure would tend to delay this peak and make jet lag worse. Bright light in the

afternoon could be expected to advance this peak (i.e., back toward Chicago nighttime). [http://www.nmjc.org/aoard/bylyons.html]

Another way to look at the use of bright light to move the melatonin peak around is to imagine a long, skinny balloon that is inflated just in the middle and still skinny at both ends. The inflated bubble represents the melatonin peak that occurs between midnight and dawn. If you squeeze the right-hand part of the bubble, it will shift left. That's like morning light pushing the peak earlier. If you squeeze the left-hand part of the bubble, it will shift right. That's like evening light pushing the melatonin peak later.

Thus, to use bright light effectively, the West Coast resident who travels to the East Coast would get up and out first thing in the morning on the East Coast so that he or she may be exposed to ½ hour or more of bright light soon after awakening. This practice, occurring just after the nocturnal melatonin peak, will help push the time of occurrence of the melatonin peak earlier (advancing the body clock), aligning it sooner with the new time zone. This approach holds for changes up to about six time zones to the east. For more than about six time zones eastward travel (you will have to experiment), substitute midday bright light exposure for early morning exposure, minimizing early bright light exposure, for the first three or four days in the new time zone, and then start early morning bright light exposure. This practice should help assure that the bright light occurs after your melatonin peak. After these three or four days, your melatonin peak should have moved early enough that the morning bright light exposure occurs after the peak.

To use bright light effectively, the East Coast resident who travels to the West Coast would emphasize bright light in the evening. The philosophy here is to extend the period of daylight experienced by the body before the nocturnal melatonin peak, to help push the melatonin peak later (delaying the body clock), aligning it sooner with the new time zone. This practice holds for changes up to about nine time zones when late daylight in the new time zone would occur at about the time of the melatonin peak you've carried with you from home. For a change of more than about nine time zones, you would expose yourself to bright light in the early and middle parts of the day, earlier than your melatonin peak.

Remember that because there are 24 time zones, if you are traveling more than 12 time zones in one direction, your time change is actually in the other direction, by the number of time zones minus 12. However, because our ability to acclimatize to new time zones is not symmetrical, a different division is used for jet lag considerations. Because we acclimatize to westward shifts in about two-thirds the time required for eastward shifts, the break comes at 10 hours when moving east: A 10-hour eastward shift or greater should be treated as a 14-hour westward shift or lesser.

What happens to aircrews who do not remain in the new time zone for more than a day or two and then return to the home time zone? What happens if you do not remain in the new time zone for more than a day or two and then continue the trip in the same or opposite direction? Frankly, we are still learning about these situations. We know that the various circadian rhythms in the body will tend to flatten out, becoming less pronounced in amplitude. We also know that the various circadian rhythms within the body may disconnect from one another. However, the prevailing, useful piece of knowledge for you is that the rhythms are still trying to express themselves in the body's functions and that you can help them.

To do this, you can keep track of your accelerated circadian adjustments as you travel and use the knowledge to your advantage. Earlier, I presented two ways to identify your physical and mental low point for your desired, normal waking and sleeping schedule at home. Alternatively, you may assume that the low point occurs at about 3 A.M. Use the following numbers to determine the accelerated changes it undergoes as you travel:

- Recall that for each 24-hour period that you spend in time zones west of home, your rhythm will probably delay (move later) by 1 hour, toward the local time. If you feel that you have been successful in using the bright light method to accelerate your body's change, assume a delay of 2 hours for that 24-hour period.

- Also, recall that for each 36-hour period that you spend in time zones east of home, your rhythm will probably advance (move earlier) by 1 hour toward the local time. If you feel that you have been successful in using the bright light method to accelerate your body's change, assume a delay of 2 hours for that 36-hour period.

You then have useful information that you can use when you consider the applications of additional melatonin and bright light to speed up your body clock's adjustment to a new time zone.

7

Cumulative Fatigue: It All Adds Up

When recovery between workdays or CDPs is absent or incomplete, cumulative fatigue will be the result. Cumulative fatigue is the fatigue that builds up across workdays or CDPs because of inadequate rest between the workdays and CDPs. This is true not only for your major sleep periods during layovers but also for days off spent at home. The development of cumulative fatigue depends on many of the factors that I have described already, the first being acute fatigue. Acute fatigue develops within one workday or CDP. With inadequate recovery between CDPs, some degree of the fatigue generated in one workday or CDP is carried over to the next workday or CDP. In this subsequent workday or CDP, the development of acute fatigue builds on top of the fatigue that has been carried forward. They are additive.

    You know now that the main factor in the development of sleepiness, a component of acute fatigue, is the prior length of wakefulness. You should also recall that other factors contribute to acute fatigue. The mental and physical work demanded of an aircrew member is a stress, to which you respond with some evidence of stress and strain. There are physiological costs, metabolic in nature, associated with physical effort. Similarly, there are psychological costs associated with mental effort. Cockpit crews face one particularly difficult type of acute, mental fatigue caused by the need to monitor cockpit automation. This task is the most likely cockpit task during which there may be a mental lapse or an involuntary sleep onset.

You know that sleep is a complex phenomenon, generated actively by the brain, and that many environmental and behavioral factors may conspire to prevent us from obtaining adequate sleep, even when we are very fatigued. A recent sleep poll suggested that inadequate sleep is a general problem in our society. These observations underscore the need for you to consider carefully the amounts and quality of sleep that you need and that you get and whether your sleep is adequate to allow recovery between workdays or CDPs. If it is not, you should change your sleep habits and be sensitive to the performance impairment associated with the cumulative fatigue that occurs with inadequate recovery sleep.

One recovery problem I've encountered centers around the effective use of time off for sleeping during a layover. We saw this during an extensive investigation of fatigue in 80 long-haul commercial truck drivers.[22] We found that although adequate sleep time was allowed, people may not take advantage of the opportunity. In the report (pages 5–18), we noted that the drivers had about 8.7 hours of time off at their sleep site, after the research time needed at the site had been subtracted from total time off. However, they tended to spend only about 5 hours in bed within that 8.7 hours. I can shave, shower, and eat two meals in less than the 3.7-hour difference between the 8.7 hours allowed and the 5 hours spent in bed. Why didn't these drivers take advantage of the time off to spend more time in bed and, presumably, sleep more? We don't know the answer. It could have been sleep biology in some cases, such as trying to sleep during the day. In many cases, it was not: They failed to sleep at night.

Whatever the causes, commercial aircrews and general aviation pilots should guard against this kind of behavior. It's possible that the same reasons that caused the truck drivers to avoid sleeping when sleep time was available could cause aircrews to do the same thing. The adage for aircrews exposed to jet lag, irregular schedules, and/or night work is to sleep whenever possible during layovers. The failure to do so can lead to a fatal accident, as it did for this 2400-hour commercial pilot on an air-taxi flight[23]:

> The pilot was cleared for a localizer approach by Atlanta Center & told to maintain 5,000 ft until crossing the final approach fix (FAF). Normal altitude at the FAF was 2,700 ft. The pilot was unable to land from this approach & performed a missed approach. He was handed off to Chattanooga Approach, then was cleared to cross the FAF at 3,000 ft & perform another localizer approach. About 1 mile from the FAF, the pilot was told to change to the airport advisory frequency. The pilot acknowledged, then there was no further communication with the aircraft. A short time later, witnesses heard the aircraft crash near the approach end of the runway. Examination of the crash site showed the aircraft had touched down in a grass area about 1,100 ft from the end of the runway, while on the localizer. Propeller slash marks

showed both engines were operating at approach power & the aircraft was at approach speed. No evidence of precrash mechanical failure or malfunction of the aircraft structure, flight controls, systems, engines, or propellers was found. The 0621 weather was in part: 300 ft overcast & ½ mile visibility with fog. Minimum descent altitude (MDA) for the localizer approach was 1,180 ft MSL; airport elevation was 710 ft. The pilot had flown 8 flight hrs, was on duty for 13.6 hrs the day before the accident, was off duty for about 6 hrs, & had about 4 hrs of sleep before the accident flight.

The probable cause cited by the NTSB was[23]:

> The pilot's improper IFR procedure, by failing to maintain the minimum descent altitude (MDA) during the ILS localizer approach, until the runway environment was in sight, which resulted in a collision with terrain short of the runway. Factors relating to the accident were: darkness, low ceiling, fog, pilot fatigue, and improper scheduling by the aircraft operator.

This was a difficult approach, but this was also an experienced instrument pilot. Adequate sleep would probably have made a lot of difference in the fate of this flight.

One beneficial change to your sleep habits may be the addition of a nap at home, during a layover, or, someday for commercial crews, while airborne. Naps can be very helpful in offsetting the potential for cumulative fatigue. However, one very important aspect of nap planning, especially when cockpit naps are allowed, is the paradox of sleep inertia, when your performance is worse after awakening than it was before you went to sleep. You have the knowledge now to support the strategic planning of naps. The thought may have occurred to you, when reading about midafternoon naps that they sound a lot like siestas. Good thinking. They are just that.

When Western Europe and the United States underwent the Industrial Revolution, we moved away from the siesta concept. Manufacturing production occupied 12 hours a day, straight through. (By the way, our later adoption of the 8-hour workday was dependent to a large degree on the sharp increase in human error rates measured during the last 4 hours of the 12-hour workday.) Those societies, most others on earth, who embrace siestas, are expressing a better understanding of fundamental human biology than we demonstrate as industrial leaders. Growing up in the southwestern United States, I had always associated siestas solely with the Mexican, South American, and Spanish cultures. Much to my surprise, I found that the siesta culture was practiced in Chinese society in Taiwan when I lived there 30 years ago. It was then I realized that we were probably missing out on a good thing, and we are.

As you can now surmise, Bernie Webb's theory about one of the benefits of sleeping during the hours of darkness, that we are best served by quiescence during the night because attempts to hunt or forage at night could lead easily to injury or death, may apply to the siesta. This is a period of the day during which error and accident risk are obviously great. Why try to work in the midafternoon and work poorly?

Thus, when strategizing about naps during your days off, please include the venerable siesta in your plans. How should you do that? If you are tired and in need of recovery sleep you may want to add a siesta nap of 2 or 3 hours to the 8 hours you get at home at night. Once you have recovered, you could eliminate the siesta. Alternatively, you could keep it going, getting at least 6 hours of sleep at night and the rest of your 8 hours during your siesta. For commercial aircrews one advantage of keeping the siesta in place is that, if your first trip after time off involves night work, you would be prepared to take a good siesta on the afternoon before the first night of work. This practice can help sustain your wakefulness and performance through that first night.

It's bad enough that acute fatigue may interact with the two-peak daily pattern of errors, in which mistakes and slips are most likely to occur during the predawn and midafternoon periods. Now, consider the amplification of the two-peak pattern by the additive nature of cumulative fatigue. Imagine how poorly you function at 4 A.M. during the first of several sequential nights of work. Now, imagine how much more poorly you function at 4 A.M. the second night after sleeping poorly during the intervening day. Finally, there is the problem of jet lag. When your SCN pacemaker is no longer aligned with the local daylight-darkness cycle, your ability to generate sleep at the times allotted for sleep may be severely impaired. Obviously, without adequate recovery sleep, cumulative fatigue is bound to occur.

I have had two experiences recently that have illustrated the nature, development, and measurement of cumulative fatigue. One was my baseline study, reported in 1998, of fatigue among crew members on U.S. Coast Guard cutters.[24] The other was a consultation to the National Interagency Fire Center concerning fatigue that developed among firefighter aircrews during the extra-long wildfire season of the year 2000.[25]

I equate a shipboard study to aircrew duties because the standard maritime watch schedule provides a reasonably good model of what aircrews must contend with. First, there is the requirement that the crew members stand watch 8 hours out of every 24. This is similar to the

8-hour flight time limit in commercial aviation. Second, there is the problem of obtaining 8 hours of sleep in an 8-hours-off period. Conversely, there is the potential for a long duty day, similar to commercial aviation operations. After the 8-hour period used for sleeping, a watchstander would be on watch for 4 hours, perform ship's work for 8 hours, and stand watch again for 4 hours. This is a 16-hour duty day, punctuated by meals. Granted, there are some differences. The main one is that the two 4-hour periods of watch standing are separated by 8 hours of other work. Although the watch-standing periods include those duties most similar to cockpit duties in aviation, the overall work demand is quite similar to that demanded of a commercial aircrew.

During the Coast Guard study, I needed to deal with the covert nature of fatigue and yet measure the development of cumulative fatigue. I used several approaches to capturing evidence of cumulative fatigue. One was to have the crew members provide multiple daily subjective estimates of sleepiness. I calculated the straight-line change in sleepiness reports across days (the linear regression line for you number crunchers) and examined that line for evidence of a progressive development of cumulative fatigue. If there was a general increase in reported sleepiness across days, I concluded that fatigue was accumulating across days. It's difficult for people to draw free hand the best-fit line through a cluster of dots; we tend to slant the line improperly. When the data have a good statistical distribution, the mathematical procedure does a much better job of drawing the best-fit line. Similarly, it's difficult for us to perceive a slow change in our own average fatigue from day to day because our fatigue level varies from hour to hour and task to task. A subtle, gradual, day-to-day change in sleepiness and fatigue is revealed more clearly with a mathematical procedure than by our own rough estimates. This is one of the reasons that I say that many signs of fatigue are covert. We have to tease the effects out of "noisy" data that do not reveal the effects on the surface.

You can follow the procedure I outlined if you know how to use a spreadsheet. You can write down your sleepiness rating at given times of day and night for several days and then use the spreadsheet's automatic line-fitting function to see if there was a gradual change across days. However, it would probably be easier for you to track how much you sleep and to try to draw some conclusions from that information.

Instances of recovery sleep can reflect the impact of cumulative or acute fatigue or both. To determine whether recovery sleep was present, I subtracted the crew members' reported ideal sleep lengths from their reported times in bed across a number of 24-hour periods. When the sleep length for a 24-hour period was shorter than a crew member's estimate of his or her ideal sleep length, I assumed that a sleep debt had been incurred. When the sleep length exceeded a crew member's estimate of his or her ideal sleep length, I assumed that recovery sleep had occurred. For example, a reported time in bed of 8 hours during the 24-hour period, minus an ideal sleep length of 7.5 hours, would indicate 30 minutes of recovery sleep. (This whole approach tends to overestimate the amount of time spent asleep because less time is spent sleeping than is spent in bed. However, the fatigued, normal individual will sleep about 98 percent of the time spent in bed. The approach would have been stronger if I could have used EEG to quantify the crew members' "ideal" and day-to-day amounts of sleep, but that was not possible.)

To determine whether cumulative fatigue had built up, I looked at the pattern of sleep debt and recovery sleep across days. An obvious pattern of sleep debt built up across days when the cutter was underway, and there was an obvious pattern of recovery sleep when it tied up at a dock. The apparent reason for this pattern was that all watches must be stood while a ship is underway but not when it is tied up. When a ship is tied up, skeleton crews take care of it. When watches were being stood on the cutters, the crew members had inadequate time to sleep enough. They made up for it when the ship was dockside.

This pattern parallels the experiences of commercial aircrews. While on a trip, flight schedules, time zone changes, local daylight and activities in other time zones, and the body clock may conspire to cause sleep debt. Upon returning home, you may experience recovery sleep.

With some experimentation and insight you should be able to determine how much sleep you need to be optimally effective each day. For the average person, that would probably be around 7 or 8 hours. Then, keep a diary across days that shows how much you sleep each day and what the difference is from your optimal sleep amount. Use quarter-hour increments for all of this; it's difficult to be more accurate. When you travel west, your day will be extended beyond 24 hours by time zone changes. When you travel east, your day will be shortened to less than 24 hours by time zone changes. Make notes of these time zone effects to help you understand why you sleep more or less.

Look at your pattern of sleep debt and sleep recovery days. You may have been aware of making up sleep after trips or periods of excessive demands at work or at home, but this procedure should give a better handle on what you need to do to recover from sleep debt. Do not be alarmed if you see that you do not generate the same number of hours of recovery sleep as your sleep debt.

For reasons that remain obscure, we do not have to generate the same numbers of recovery sleep hours as sleep debt hours. Recovery sleep seems to occur at a ratio of about one-half or one-third of sleep debt. That is, if you have built up a sleep debt of 6 hours, you will probably find that 2 or 3 hours of extra, good quality sleep will make up for it. By good quality, I mean that you are sleeping during your body clock's nighttime and in a place where you can sleep soundly and uninterrupted for 9 or 10 hours. One thing to watch for in this process is the build up of sleep debt before flights or trips. Try to avoid that.

Chronic Fatigue

I found that another kind of cumulative fatigue developed among firefighter aircrews during the wildfire season of 2000. It was associated with mild depression. Dealing with difficult, ill-defined, complex problems is mentally fatiguing, *especially* if no obvious solution is forthcoming. Guilt and anxiety over one's limited abilities to address a problem may lead to mild depression, a symptom of which is fatigue. When one invests personally and extremely in an effort and then fails to meet expected goals, feelings of guilt and anxiety may result, leading to mild depression and fatigue. For example, the agency employees who left work and family behind throughout the fire season, only to realize that the original goal, to put the fires out, could not be met, questioned the value of their investments of time, energy, and emotion. This questioning appeared to lead to feelings of guilt and anxiety, contributing to mild depression and fatigue. Of course, just the fact that they were coping only marginally with the year's extreme demands for fire management probably led to feelings of guilt and anxiety and to mild depression and more fatigue.

Nearly all pilots and fire managers I spoke with expressed frustration at their inability to get the fires out that season. Of course, because of the extent of the fires, firefighting priorities changed from getting the fires out to safety, the protection of life, and the selective protection of high-value structures, and not even all structures could be protected, much less defended. However, this official change in priorities did not remove the frustration associated with the situation. Firefighters expect to put fires out. Thus, in some cases, guilt and anxiety leading to mild depression and fatigue occurred. This frustration appeared to be a major contributor to cumulative fatigue in this difficult fire season. Even days off could not diminish the level of frustration.

Discussions with firefighters emphasized for me the interactions among job demands, job satisfaction, attitude, and fatigue. Commercial aircrews experience several of the problems I described for firefighters: investing personally and extremely in an effort, leaving family behind, and failing to meet expected goals. All aircrew members have invested heavily in time, effort, and, often, money to enter the profession. Just making it through initial training is demanding. Subsequent on-the-job training in the cockpit and cabin and repeated flight checks require continual skill development and maintenance. For commercial crews, family is left behind for days on end during trips. When we're unattached, this may not be viewed as a problem. For those whose significant other works a 9-to-5 job or who have kids in school, this irregular work schedule and absence from the community can be a real problem. Increasingly, commercial aircrews are frustrated by their inability to accomplish one of their main goals—moving travelers in a timely manner. Traffic density and control and weather have taken their tolls in recent years on the timeliness of scheduled flights. (Yes, as a traveler I've had many flights canceled in the last few years.)

There is probably no cure-all for dealing with the effects of these problems on your performance. There are simply too many variables and too many individual differences here for one approach. In fact, some aircrew members are pretty much unconcerned about all of these issues. However, for those who are affected, your main lesson should be to realize that these problems can, literally, "get you down" in terms of performance. They can be fatiguing through the links of guilt and anxiety that I described, above.

When I went through USAF pilot training in 1966–1967 at Webb Air Force Base in Big Spring, Texas, there was an official, organized attempt to indoctrinate the pilot-trainees' wives (women were not military pilots then) as important aids to flight safety. They were told how distractions caused by family life could lead to their husbands making mistakes in the cockpit and crashing. Thus, the wives were supposed to "take care" of all of the minor housekeeping irritants that arise in life and not bother their husbands with any issues that would cause them anxiety. This would have been a sweet deal for us guys. Unfortunately, the wives' group attitude was, "Suck it up you wimps and be men!" Ah well. Such is life. However, you can see the point that the Air Force was trying to make at the time. Anxiety can be a problem. They termed it a distracter. You can also easily see that the mental effort associated with guilt and anxiety can lead to psychological cost and fatigue. It's there. We just need to recognize it, keep it in mind, and deal with it.

One of the ironies of dealing with anxiety-producing problems at home is the time required to fix them. Often, the time at home between trips is inadequate for making much progress, and your recovery from cumulative fatigue is impaired. In many cases, travel home diminished the value of days off for firefighters, contributing to cumulative fatigue. They reported that, among other things, work and family demands contributed to cumulative fatigue during days off. In addition, for the firefighters, just as for aircrews, uncertainties about subsequent assignments after days off were reported as frustrating. When assignments were changed repeatedly and/or at the last moment, crews and staff reported feeling as though they were being "jerked around." Most of us are familiar with the statement, "I'm tired of being jerked around." For the reasons stated above, this is literally true: We do get tired.

From the shift-work research literature, most of it from Europe, we have learned that two important components of job satisfaction for shift-workers are schedule predictability and scheduling equity. The firefighters were speaking of the latter, predictability. It's good to know whether we'll be working or not, or in town or not, 6 or 12 or 18 weeks from now. Obviously, this is why one of the attractive prospects of seniority in the airlines is the ability to bid trips successfully and, thus, plan your life. In a way, this works well. Younger people probably recovery from fatigue more quickly than older folks do. Thus, those who have less seniority usually bear the burden of low schedule predictability.

Scheduling equity is another factor in satisfaction. Are you getting about the same number and quality of trips as others with your qualifications? Scheduling equity has long been a problem in commercial and military aviation. There are so many scheduling variables that it is difficult to write serially structured computer code to encompass the huge matrix of interactions among all of the variables. Some of the variables might include crew position, seniority, qualification, currency, total time flown recently, availability, aircraft type, and aircraft availability. If there were only three possible categories in each of those eight variables, the matrix to be considered would have $3^8 = 6,561$ cells.

Try plugging all of your personnel and aircraft into that matrix and the equity problem is tough for both humans and computers. Now, with the advent of theories of and programming methods for fuzzy logic and neural networks, the scheduling equity problem can begin to be addressed adequately with computers.

Underestimation

Finally, you need to be sensitive to the fact that we tend to underestimate our degree of fatigue. For example, a member of an area fire management team described the following experience. He felt as though he was not very fatigued at the end of 21 days of work. However, on both of his days off, he slept in unexpectedly and, also unexpectedly, napped on both days. He reported that he "doesn't nap." The amount that he slept on his days off surprised him, compared to his expectations at the end of his 21-day work period. One must also wonder if his recovery was adequate after only 2 days off, because his sleep patterns had not returned to normal.

This underestimation effect is also present with alcohol. Most of us have heard, or said, "I'm not drunk. I can drive." In addition, most of us have been told repeatedly over the last 10 or 20 years to get the car keys away from the folks who are saying those things. A recent safety campaign along these lines was called "Carpe Keyem," seize the keys. When we ask sleep-deprived subjects or subjects with 0.05 and 0.10 blood alcohol content to estimate how well they will perform various cognitive tasks in the laboratory, both sets of subjects overestimate their capabilities. The lesson? Once you realize that you are fatigued, it's probably worse than you suspect. Double- and triple-check all of your work.

8

Can the Pharmacy Help?

As we all know, the abuse of drugs and alcohol may lead to serious accidents. For example, the Aircraft Owners and Pilots Association's 1999 Nall Report noted that (for the last year of available data)[26]

> In 1997, 17 general aviation aircraft accidents involved drug or alcohol abuse. Twelve of these were fatal, killing 22 people. Alcohol was involved in seven accidents, including four fatal accidents, accounting for seven deaths. Illegal drugs figured in four accidents, three of them fatal, resulting in six deaths. The other six accidents, including five fatal accidents claiming nine lives, involved improper use of legal medications.

However, on the flip side, one of the strategies used to deal successfully with fatigue can be pharmacological. A convenient way to divide and conquer our consideration of this strategy is in terms of "go" (alerting) and "no-go" (sleep-inducing) treatments and in terms of prescription and nonprescription treatments. Two caveats. First, my descriptions of the newer prescription drugs listed here will become dated within a year or two as new research is published. Second, your employer and/or your regulator, such as the Federal Aviation Administration, may not approve of your use of these countermeasures in association with your flying duties.

Go Treatments

Go treatments are potentially available to sustain or elevate your level of performance when you must fly fatigued.

Nonprescription options

Caffeine is probably the most common nonprescription alerting treatment used in aviation. It can be taken in liquid (coffee, tea, colas) or pill forms. A cup of coffee provides about 100 milligrams of caffeine, and SmithKline Beecham's Vivarin and Bristol-Myers Squibb's NoDoz pills each contain 200 milligrams of caffeine. Caffeine peaks in the blood stream in about 45 minutes and then falls to half that value in another 3 to 4 hours (its half-life). Used only when needed and with reasonably good sleep beforehand, caffeine can be a highly effective fatigue countermeasure that sustains performance. However, if too much is used chronically, its alerting benefits are lost through habituation. Chronic overuse can also cause dehydration, nervousness, irritability, and sleeplessness (see figure on next page). Of course, caffeine affects people's sleep differently. For some, even small amounts early in the day can cause problems sleeping at night. For others, caffeine has no apparent detrimental effect on subsequent sleep. If caffeine does affect your sleep, it's best to avoid coffee, tea, or caffeinated soft drinks within 6 or 7 hours of your bedtime.

Can the Pharmacy Help? 129

> The following are ten clues, anonymously posted to the Internet and edited, that you drink too much coffee:
>
> 10. Juan Valdez names his donkey after you.
> 9. You get a speeding ticket when you are parked.
> 8. You grind your coffee beans in your mouth.
> 7. You can jump-start your car without a battery.
> 6. Starbuck's owns your mortgage.
> 5. You're so wired, you pick up National Public Radio.
> 4. Your life's goal is to amount to a hill of beans.
> 3. Instant coffee takes too long.
> 2. You want to be cremated so you can spend eternity in a coffee can.
> 1. You sleep with your eyes open.

Certain over-the-counter (OTC) pain relievers at your local pharmacy also contain caffeine, so check the labels. The "active ingredients" part of OTC drug labels will tell you the caffeine dose in milligrams. A brief period of study at my local pharmacy showed me that

- Aspirin Free Excedrin contains acetaminophen and 65 milligrams of caffeine.
- Excedrine Migraine contains acetaminophen, aspirin, and 65 milligrams of caffeine.
- Bayer's Vanquish Pain Reliever contains acetaminophen, aspirin, and 35 milligrams of caffeine.
- Goody's Headache Powder contains acetaminophen, aspirin, and 32.5 milligrams of caffeine.

Ingredients in cold medicines that are available over the counter, such as ephedrine, pseudoephedrine, and phenylpropanolamine, can also act as stimulants. Ephedrine triggers the release of adrenaline, the fight-or-flight hormone, in the body. Weight loss aids can also contain stimulants. For example, TwinLabs' Ripped Fuel contains 334 milligrams of MaHuang extract, standardized for 6 percent ephedrine, and 910 milligrams of Guarana extract, standardized for 22 percent caffeine. From their numbers, I assume that you would get a dose of 20 milligrams (0.06×334) of ephedrine and 200 milligrams (0.22×910) of caffeine in one capsule. The FDA suggested in 1997 that diet pills be limited to doses of 8 milligrams of ephedrine in a 6-hour period or a total daily intake of 24 milligrams. That suggestion was withdrawn, and they are now seeking public comment and reports of adverse reactions (see *vm.cfsan.fda.gov* on the Internet).

Nicotine is another common nonprescription alerting treatment used in aviation. Unfortunately, although it is a stimulant, it turns out that it is not very effective in offsetting the performance impairments caused by lack of sleep.

Prescription options

Early in 1968, I was issued a pack of six *dextroamphetamine* (Dexedrine, or Dex) tablets. Our C-130E squadron (the 346th Tactical Airlift Squadron, the "Black Knights") was moving two-thirds of the way around the world, from Mildenhall in England to Clark Air Base in the Philippines, to help deal with the Tet Offensive in South Viet Nam. We were allowed to use the tablets if our crew duty day was extended past 18 hours and the crew was not augmented with additional pilots. If we used the Dex, we were to report to a flight surgeon before our next sortie for medical clearance.

Today, Dex is still available to U.S. Air Force Flight Surgeons in our Air Combat Command, but only for use in aircraft with single pilots for sorties longer than 8 hours. It has been used successfully in recent conflicts. F-111B aircrews who used 5 milligrams Dexedrine during the Air Force strike on Libya experienced positive effects in terms of overcoming the fatigue associated with the mission and the sleep deprivation that occurred during preparation for the mission. F-15C pilots flying 6- to 11-hour combat air patrol missions during Operation Desert Shield/Storm were sleep deprived and suffered from circadian rhythm disruptions. Each pilot was issued five or six 5-milligram tablets at the beginning of a mission and was directed to self-administer one tablet every 2 to 4 hours as needed to maintain alertness until landing. There were no reported adverse effects, even in pilots who took 10 milligrams at a time, and none reported a need to continue the drug after typical work/sleep schedules were reinstated.

Dr. John Caldwell and his colleagues at the U.S. Army Aeromedical Research Laboratory at Ft. Rucker have found that several repeated 10-milligrams doses of Dex improved the performance of sleep-deprived helicopter pilots flying UH-60 simulator missions and found no clinically significant behavioral or physiological effects in any of the pilots. In the U.S. Air Force, use for two-pilot, long-duration bombing missions may be initiated in 2001. These missions in the B-1 and B-2 bombers may last up to 36 hours, with multiple refuelings.

One of the common worries with the use of Dex is addiction. However, the extremely limited use of Dex in military aviation has never to my knowledge generated a problem with addiction. A more worrisome problem is the feeling of euphoria that accompanies the stimulant properties of Dex. There is concern that a pilot could become overconfident during a critical point in a mission.

I am not aware of any such occurrence, but it is cause for concern. Finding a replacement or an alternative for Dex would be useful. The main candidate at this time is modafinil.

Data from military personnel have suggested that *modafinil* (Provigil) is effective in sustaining alertness but, unlike Dex, has the additional benefits of not producing sleep difficulty when sleep time is available. Modafinil has been found to produce fewer of the side effects associated typically with amphetamines, such as euphoria, high abuse potential, and tremor. One hundred milligrams of modafinil given every 8 hours during 64 hours of sleep deprivation is enough to maintain, but not improve, mental performance. A 200-milligram dose of modafinil can improve physical ability in nonfatigued persons. The 300-milligrams dose used in most lab studies of modafinil compares favorably with 10 milligrams of amphetamine in ameliorating the effects of sleep deprivation. Thus, modafinil appears to have good fatigue-reducing abilities and minimal side effects. It has not yet been tested in operational simulations.

A recent study in Dr. Caldwell's lab, with Army pilots operating the full-motion UH-1 helicopter simulator, supported the usefulness of modafinil as a go pill that may be of use in military operations. They used three doses of 200 milligrams each across one night of simulated flight operations. However, some of the pilots in the study reported some vertigo. Thus, an Air Force study is underway, as I write this, to look solely at the effects of single doses of 200 and 400 milligrams of modafinil on the vestibular system.

No-Go Treatments

No-go treatments are potentially available to help you sleep when night work, circadian rhythms, or transmeridian travel impairs your brain's ability to generate adequately restful sleep.

Nonprescription options

Obviously, aircrew must comply with federal regulations about avoiding *alcohol* consumption preceding flight. However, questions probably remain about the effects of alcohol on sleep obtained during days off. Alcohol may make you feel drowsy, but its ultimate effect is that you'll sleep less soundly and awake more tired as a result of drinking. Both deep sleep and dreaming sleep will be suppressed, and you will experience early-morning awakenings, perhaps with a hangover. A drink before dinner or a glass of wine with dinner probably won't make too much difference. However, avoid moderate or greater alcohol consumption within 6 hours of bedtime if you want to sleep well. An extract from the National Institute on Alcohol Abuse and Alcoholism *Alcohol Alert*[27] describes the problem quite well:

> Alcohol consumed at bedtime, after an initial stimulating effect, may decrease the time required to fall asleep. Because of alcohol's sedating effect, many people with insomnia consume alcohol to promote sleep. However, alcohol consumed within an hour of bedtime appears to disrupt the second half of the sleep period. The subject may sleep fitfully during the second half of sleep, awakening from dreams and returning to sleep with difficulty. With continued consumption just before bedtime, alcohol's sleep-inducing effect may decrease, while its disruptive effects continue or increase. This sleep disruption may lead to daytime fatigue and sleepiness.

Alcoholic beverages are often consumed in the late afternoon (e.g., at "happy hour" or with dinner) without further consumption before bedtime. Studies show that a moderate dose of alcohol consumed as much as 6 hours before bedtime can increase wakefulness during the second half of sleep. By the time this effect occurs, the dose of alcohol consumed earlier has already been eliminated from the body, suggesting a relatively long-lasting change in the body's mechanisms of sleep regulation.

The adverse effects of sleep deprivation are increased following alcohol consumption. Subjects administered low doses of alcohol following a night of reduced sleep perform poorly in a driving simulator, even with no alcohol left in the body. Reduced alertness may potentially increase alcohol's sedating effect in situations such as rotating sleep-wake schedules (e.g., shift work) and rapid travel across multiple time zones (i.e., jet lag). A person may not recognize the extent of sleep disturbance that occurs under these circumstances, increasing the danger that sleepiness and alcohol consumption will co-occur.

By the way, if the question has occurred to you—yes, there is an approximate equivalency between sleep deprivation effects and alcohol impairment. Presently, research results suggest that 17 hours of continuous wakefulness is the equivalent of about 0.05 blood alcohol content (BAC) in terms of performance impairment. In the laboratory, 0.05 BAC is the point at which we can start to make statistically reliable measurements of people's performance impairment. This is not to say that their performance is unimpaired between 0 and 0.05 BAC. Human

Can the Pharmacy Help? 135

performance varies a lot when alcohol is in the body. That variability prevents us from making reliable statistical statements about performance impairment in that range.

In fact, fatigue and alcohol consumption are analogous to a limited degree. In both conditions, performance variability increases gradually and average performance decreases gradually. As we remain awake beyond some number of hours, performance variability increases gradually and average performance decreases gradually. As our blood alcohol increases beyond zero (after the first 30 minutes or so), performance variability increases gradually and average performance decreases gradually. The use of this analogy is becoming more common, and it seems to be a reasonable analogy.

As with the go treatments, some no-go treatments are available as OTC products at your local pharmacy. Pfizer's Unisom uses the antihistamine, doxylamine, as a sleep aid, and Tylenol's Simply Sleep, Block's Nytol and Sominex use the antihistamine, diphenhydramine. Antihistamines do, indeed, make you sleepy. In fact, that's one of the precautions noted for antihistamines on the National Library of Medicine's (NLM) web site (*www.nlm.nih.gov*):

> [Antihistamines] may cause some people to become drowsy or less alert than they are normally. Even if taken at bedtime, it may cause some people to feel drowsy or less alert on arising. . . . Make sure you know how you react to the antihistamine you are taking before you drive, use machines, or do anything else that could be dangerous if you are not alert.

The NLM also notes that

> Antihistamines will add to the effects of alcohol and other CNS depressants. . . . Some examples

of CNS depressants are sedatives, tranquilizers, or sleeping medicine; prescription pain medicine or narcotics; barbiturates; medicine for seizures; muscle relaxants; or anesthetics, including some dental anesthetics. *Check with your doctor before taking any of the above while you are using this medicine.*

Melatonin is a hormone that is produced by the brain (actually, the pineal gland) that helps induce sleep. Melatonin supplements are advertised widely as sleep aids and melatonin has received widespread public support, but not widespread scientific support, as a safe and nonprescription means to induce sleepiness. In one study, a dose of 10 milligrams was just as effective as a 40-milligram dose in promoting sleep.

Although melatonin may not be as effective as prescription drugs (below) in inducing sleep, it has the distinct advantage of not promoting sleep by intoxication, which would preclude aircrews from duty until drug washout. However, questions about long-term safety have not yet been answered; this is, after all, a hormone produced in the body already. The U.S. Air Force Surgeon General has directed that USAF aircrews not use melatonin until the long-term health benefits are better understood. The position of the National Sleep Foundation about melatonin seems to be similar in nature.

Prescription options

Prescription sleep medications do not cure sleep problems. However, they may be helpful.

Triazolam (Halcion), a benzodiazepine, has been prescribed widely as a sleep aid for many years. Its use by aircrew is discouraged. Among some other problems, there appears to be a strong drug hangover effect that persists well after awakening from sleep that has been induced by triazolam.

Temazepam (Restoril), a newer benzodiazepine, appears to have minimal hangover effects. However, as noted on the Internet by Phillip W. Long, M.D.,[28] "The most common adverse reactions reported after administration of temazepam and other drugs of this class are dizziness, lethargy, and drowsiness. Confusion, euphoria, staggering, ataxia, and falling are commonly encountered." There is some concern about recovering from these effects after sleep, before flying. On the other hand, temazepam was used, with glowing reports of success from combat aircrews and flight surgeons, in the Falklands and in Desert Storm.

Zolpidem (Ambien), an imidazopyridine, has a shorter duration of action and less severe side effects than temazepam. Studies to date suggest that zolpidem produces no rebound or withdrawal effects and patients have experienced good daytime alertness after 20-milligram oral doses at night. Peak plasma concentrations are reached about 45 minutes after ingestion. This profile makes zolpidem the best pharmaceutical alternative at present for use as an aircrew sleep aid. In fact, it has replaced temazepam (which had replaced triazolam) in military formularies used by flight surgeons for aircrews. However, zolpidem can produce some benzodiazepine-like side effects: drug hangover and anterograde amnesia. Aircrews may be impaired until the compound wears off.

Zaleplon (Sonata), a pyrazolopyrimidine, appears to have sleep-inducing properties similar to zolpidem and, perhaps, even fewer side effects.

Other prescription drugs can disrupt sleep. Bronchodilators may contain some of the same ingredients as cold preparations. Diuretics taken too close to bedtime can cause you to use the bathroom several times during a sleep period. Antidepressants such as Prozac and overmedication with thyroid drugs could cause sleeplessness, as may some antimigraine drugs. If you do take a drug on a regular basis and have difficulty sleeping, consult with your physician about whether the drug could be responsible for some of the sleep disturbance. Perhaps the dosage and timing of the drug can be modified.

Taking prescription drugs and then flying can certainly lead to problems. Consider the fate of this 309-hour commercial pilot at midnight in the middle of July[29]:

> The pilot departed on a VFR flight and did not return to the departure airport. The airplane wreckage was located the following day. During examination of the airframe, flight controls, engine assembly, and accessories, no precrash failure or malfunction was found. The investigation revealed that the pilot had been admitted to a local hospital 2 days before the accident for gastroenteritis. His illness resulted in dehydration, disrupted sleep pattern, and inadequate nutrition. The prescription drug phenergan (promethazine) was prescribed for the pilot. Toxicology tests of the pilot's blood showed 0.034 μg/ml promethazine, and promethazine was detected in the pilot's urine. Promethazine produces side effects of drowsiness.

To restate that which I think should be obvious to all: Flying while taking prescription drugs or while self-medicating can be fatal. Either abstain from flying or consult your physician for guidance. The risks and benefits for aircrews of the main go and no-go treatments are summarized in the following table.

Can the Pharmacy Help?

Treatment	Benefits	Risks
Go		
Nonprescription		
Caffeine	Highly effective stimulant	Moderately addictive; chronic dehydration, nervousness, irritability, and sleeplessness; rapid tolerance
Prescription		
Dexedrine	Highly effective stimulant	Highly addictive; chronic nervousness, irritability, and sleeplessness; euphoria (overconfidence)
Provigil	Highly effective stimulant	Few side effects reported to date
No-Go		
Nonprescription		
Alcohol	Initial sedative	Hangover; disruption of sleep; moderately addictive; impairment of motor skills; euphoria (overconfidence); anterograde amnesia; dehydration
Antihistamines	Often an effective sleep aid	Drug hangover; dry mouth, nose, or throat; gastrointestinal upset, stomach pain, or nausea; increased appetite and weight gain; thickening of mucus
Melatonin	Often an effective sleep aid	Long-term effects on health unknown

(*continued*)

No-Go (*Continued*)

Prescription		
Halcion	Effective sleep aid	Drug hangover; slow-wave sleep suppression; dizziness, lethargy, and drowsiness; confusion, euphoria, staggering, ataxia, falling; anterograde amnesia; rapid tolerance
Restoril	Effective sleep aid	Slow-wave sleep suppression; anterograde amnesia; drug hangover; rapid tolerance
Ambien	Effective sleep aid	Anterograde amnesia; drug hangover?
Sonata	Effective sleep aid	Not yet determined

9

A Prescription for Fighting Fatigue

I hope that, by now, you have accepted the fact that night work and jet lag seem always to compromise a person's ability to acquire adequate sleep and, thus, to recover between duty periods from acute fatigue. Concomitantly, you should be sensitive to the performance impairment associated with the cumulative fatigue that occurs with inadequate recovery sleep. Acceptance of the reality of these phenomena is crucial if you are going to do something to help yourself combat fatigue.

Perhaps you have no trouble accepting these ideas. On the other hand, I have met people who refuse, even in the face of multitudinous personal testimonies and volumes of quantitative human performance data, to admit that they become fatigued like the rest of us. To my knowledge, however, none of my colleagues or I have ever collected data on people who say this. Maybe they are correct: They do not experience fatigue like the rest of us. However, if so, they must be rare because I've never encountered one among the hundreds of subjects who have volunteered for the fatigue studies with which I have been associated.

Thus, the chances are that the irregular schedules, night work, and jet lag associated with aviation will fatigue you. This being the case, the information in the preceding chapters should be of use to you. However, how do you organize all of this information into a useful set of instructions that you can apply on a regular basis to combat fatigue? The approach that I've used here is to sort the advice from the preceding chapters into three categories: what to do to get started, what you can do at home between trips, and what you can do during trips.

Getting Started

One of the first things to do is to consider carefully the amounts and quality of sleep that you need and that you get and whether your sleep is adequate to allow recovery between workdays or CDPs. If it is not, you should change your sleeping habits. I have summarized some general hints for good sleep hygiene in Appendix A.

Is your sleep adequate in general? The present guideline from sleep professionals is to get 8 hours per *night* (note that I emphasize the word, *night*, when the recovery provided by sleep is best). You may answer the question for yourself by estimating your daytime sleepiness with the Epworth Sleepiness Scale. If your score is greater than 15, you have a serious problem and may even wish to consult your physician in addition to implementing the suggestions provided here. If you are an Internet user, you may wish to read materials provided by the National Sleep Foundation (NSF) at *www.sleepfoundation.org*. You may also contact them by mail at 1522 K Street, NW, Suite 500, Washington, DC 20005.

If your Epworth score is 10 to 15, you have a moderate problem. You need to implement many of the suggestions provided here and by the NSF. If your Epworth score is less than 10, you are doing well. The suggestions provided here and by the NSF may help you fine-tune your sleep hygiene or nutrition.

To help you estimate your level of sleepiness from hour to hour, copy the Stanford Sleepiness Scale (SSS) onto a card and make it handy. For example, carry it with you in a pocket-sized appointment book. Select the number, 1 through 7, that best describes how you feel each hour. Write the number into your appointment book or some other note-taker so you can review and summarize the information later.

Learn about your body's clock. When you have been at home for enough days to feel that you are well established on your normal, desired home schedule, figure out the approximate time at which your metabolism hits its low point. Use the SSS or your body temperature. The time should be around 3 or 4 A.M.

Consider adding an afternoon siesta nap of 2 or 3 hours to the sleep you get at home at night. If you do this, get at least 6 hours of sleep at night and the rest of your 8 hours or more during your siesta. One advantage of using a siesta is that, if your next flight or trip requires night work, you would be prepared to take a good siesta on the afternoon before the first night of work. This practice can help sustain your wakefulness and performance through that first night.

If it is legal for your operation, add cockpit naps to your flight plan. I provided the design of a table for doing this, and a blank, full-page table is available in Appendix B. Be sure to allow time to recover from sleep inertia before critical events: at least 30 minutes. If you have missed all or most of a night of sleep, your sleep inertia could last ½ to 2 hours after a nap of up to 45 minutes or even 3 hours after a nap of 90 minutes. If you are sleep deprived, you may wish to avoid napping at all during the predawn hours if you must complete a critical task within a couple of hours after the nap.

Make a plan for good nutrition and stock your larder accordingly. Your plan should include adequate levels of protein, complex carbohydrates, and dietary fiber and small amounts of monosaturated fats. Remember to watch for symptoms of lactose intolerance.

Educate yourself or, for managers and safety personnel, educate aircrews and maintenance personnel about the intelligent management of their principal sleep periods and naps. Brochures are available from many sources to help with this effort.

When scheduling flights, you should change flight schedules wherever possible to deal with the phenomena of sleep and circadian rhythm biology.

- Determine those flight legs that are most vulnerable to loss of attention
- Examine the recent patterns of incidents (mistakes and slips, errors of commission and omission, and judgment errors) that occur in the different legs
- Rank these legs using a scale or set of scales that reflects each leg's vulnerability to errors and the potential for loss of life, health, and property if an attentional lapse occurs
- Avoid scheduling these critical flight legs during the predawn hours

In addition, commercial aviation schedulers should emphasize schedule predictability and scheduling equity for aircrews.

During Trips

If possible, you should not fly between midnight and dawn. Thus, first and foremost, you need to estimate when flight legs must be accomplished during those hours and plan to manage the elevated risks associated with flight during those hours. Assume that you will make a much higher number of errors of omission and commission than usual during that period. Double- and triple-check all of your work and have all crew members double- and triple-check each other's work.

As you travel east or west across time zones, your body clock will drift slowly from your home time. To estimate the drift of your body clock time during a trip across time zones,

- Subtract 1 hour from home time for each 24-hour period spent in time zones west of your home time zone
- Add 1 hour to home time for each 36-hour period spent in time zones east of your home time zone

Always remember that

- Seventeen hours of continuous *wakefulness* (not just work, but *wakefulness*) is the equivalent of about 0.05 blood alcohol content in terms of performance impairment.
- The occurrence of mental lapses in the cockpit or before flying is a signal to you that you are sleepy and less vigilant than you should be.
- Once you realize that you are fatigued, it's probably worse than you suspect.

When these fatigue indicators are present, double- and triple-check all of your work and have all crew members double- and triple-check each other's work.

Optimize your sleeping habits and accommodations during layovers. Guard against misusing your time off by not sleeping enough: Sleep whenever possible during layovers. Plan to sleep during your body clock's nighttime. Plan naps during your body clock's midafternoon slump. Use sleep masks over your eyes and earplugs, if these are reasonable options for your accommodations. Arrange to keep housekeeping away from your room during the day, if necessary.

Used only when needed and with reasonably good sleep beforehand, caffeine can be a highly effective fatigue countermeasure. However, if caffeine affects your sleep, avoid coffee, tea, or caffeinated soft drinks within 6 or 7 hours of your bedtime. Certain over-the-counter pain relievers at your local pharmacy contain caffeine, so check the labels.

Before you sleep, avoid large, high-fat, high-acid meals; sweets; and hunger. If you eat soon before you plan to sleep, emphasize grains, breads, pastas, vegetables, and/or fruits.

A drink before dinner or a glass of wine with dinner probably won't make too much difference. However, avoid moderate or greater alcohol consumption within 6 hours of bedtime if you want to sleep well.

Drink plenty of water every day: about one ounce of water for every two pounds of body weight (about 30 milliliters per kilogram of body weight). That's 50 ounces of water, or about 1.5 quarts, per 100 pounds of weight (about 1.5 liters per 50 kg of body weight).

Allow for sleep inertia before and during flight planning. If you are very fatigued when you go to sleep, your sleep inertia could last up to 3 hours. Double- and triple-check all of your work and have all crew members double- and triple-check each other's work.

To use bright light effectively to try to accelerate your body clock's adjustment to new time zones,

- When traveling east, up to about 6 time zones in one or two days, get up and out first thing in the morning to get ½ hour or more of bright light soon after awakening. For more than about six time zones, substitute midday bright light exposure for early-morning exposure, minimizing early bright light exposure.
- When traveling west, up to about nine time zones in one or two days, emphasize bright light in the evening. For a change of more than about nine time zones, expose yourself to bright light in the early and middle parts of the day, minimizing evening bright light exposure.

At Home between Trips

You need to be fully rested when you start a flight or a trip—carry no cumulative fatigue into that flight or first CDP; it will simply add to the acute fatigue you experience. Thus, when preparing for a a flight or a trip or when on a trip, you should trade awake time for sleep time whenever possible.

The present guideline from sleep professionals is to get 8 hours of sleep per night. If you are tired and in need of recovery sleep you may want to add an afternoon siesta nap of 2 or 3 hours to the 8 hours you get at home at night. Once you have recovered, you could eliminate the siesta. Alternatively, you could keep it going, getting at least 6 hours of sleep at night and the rest of your 8 hours during your siesta. One advantage for commercial aircrews of keeping the siesta in place is that, if your first trip after time off involves night work, you would be prepared to take a good siesta on the afternoon before the first night of work. This practice can help sustain your wakefulness and performance through that first night.

Remember that recovery sleep after excessive work or home demands or trips will probably occur at a ratio of about one-half or one-third of sleep debt. That is, if you have built up a sleep debt of 6 hours, you will probably find that 2 or 3 hours of extra, good quality sleep will make up for it. By good quality, I mean that you are sleeping during your body clock's nighttime and in a place where you can sleep soundly and uninterrupted for 9 or 10 hours.

Remember that guilt and anxiety can, literally, "get you down" in terms of performance. The mental effort associated with guilt and anxiety can lead to psychological cost and fatigue. It's there. Recognize it, keep it in mind, and deal with it. Try to strike a balance between time spent resolving anxiety and guilt-producing problems at home and time spent recovering from acute or cumulative fatigue.

Diagnosis: Fatigue

When I read a report about an aviation incident, I find it difficult to determine whether fatigue was truly a player in the chain of causes. Usually, I find partial information about the aircrew's work-rest schedule, about some performance error the crew has made, and about their self-report of their perceived degree of fatigue. This leaves me in the position of making incomplete inferences about the efforts expended by the crew to meet the work demand, the crew's degree of motivation, and the physiological and mental costs of the efforts. You could help the operational and aviation research communities by including the following information in reports about fatigue-induced incidents:

- The actual workday or CDP and rest schedule for the last three work-rest cycles, using specific times of day and including details of all time zone changes, when applicable
- The types and timing of physical and mental stresses experienced by the crew members across the last three work-rest cycles
- The degree of effort used by the crew members to respond to these stresses (10 percent? 100 percent?)
- Your estimates on scales of 1 to 10 of the physiological and mental costs of your efforts
- The quality of crew performance across the three cycles
- The degrees of fatigue you experienced, on a scale of 1 to 10, across the three cycles

Keep this list in mind as you contemplate the interactions of fatigue, its causes, and its effects in your life as a pilot and you'll have the "big picture" about fatigue.

Fighting Fatigue

Fatigue usually presents itself covertly as a set of symptoms that are known primarily only to you. Remember I. D. Brown's admonitions that[1]

- "Fatigue may be conceptualized as the subjective experience of individuals who are obliged . . . to continue working beyond the point at which they are confident of performing their task efficiently."
- "The main effect of fatigue" is "a progressive withdrawal of attention" from the task at hand that "may be sufficiently insidious that [operators]

are unaware of their impaired state and hence in no position to remedy it."

Fatigue-related accidents continue to pervade all modes of transportation, explaining why fatigue has been on the "top ten" wanted list of the National Transportation Safety Board for more than 10 years. Fatigue is caused by factors that occur in three major categories: acute fatigue, circadian effects, and cumulative fatigue. Acute fatigue develops within one work period, and one may recover from it during one major sleep period. Unfortunately, work and flight schedules, jet lag, and sleep biology conspire to prevent this from happening for many aircrews. Consequently, cumulative fatigue and the malaise of jet lag are all too common among flight crews.

Having read this book, you should now be aware of the many complexities that underlie the apparently simple process of acquiring enough rest to perform your job safely and effectively. You should have gained some useful knowledge and practices to use to combat the fatigue that comes naturally with the business of being an aircrew member. Use the knowledge and practices you have gained to experiment; learn what might work for you. Learning the characteristics of your personal fatigue susceptibility and determining which practices are the best fatigue fighters for you can lead to greater alertness, better work performance, and enhanced flying safety.

References

1. I. D. Brown, Driver fatigue. Human Factors 36(2): 298–314, 1994.

2. Office of Technology Assessment, Biological Rhythms: Implications for the Worker, Committee Report OTA-BA-463, U.S. Government Printing Office, 1991.

3. NTSB Aviation Accident/Incident Database, Report No. DCA93RA060, 08/18/1993.

4. Mackie, R. R., and Miller, J. C., Effects of Hours of Service, Regularity of Schedules, and Cargo Loading on Truck and Bus Driver Fatigue (HFR-TR-1765-F), Goleta, Calif., Human Factors Research, Inc., 1978.

5. Singh, L.L., Molloy, R., Parasuraman, Raja, Individual differences in monitoring failures of automation. J. General Psychology 120(3):357–373, 1993.

6. Miller, J. C., Smith, M. L., and McCauly, M. E., Crew Fatigue and Performance on U.S. Coast Guard Cutters, Report CG-D-10-99, U.S. Coast Guard Research and Development Center, Groton, Conn., 1999.

7. NTSB Aviation Accident/Incident Database, Report No. NYC91FA001, 10/02/1990.

8. NTSB Aviation Accident/Incident Database, Report No. FTW98FA287, 06/28/1998.

9. NTSB Aviation Accident/Incident Database, Report No. CHI97FA027, 11/15/1996.

10. Mitler, M. M. and Miller, J. C., Methods of testing for sleepiness. Behavioral Medicine 21:171-183, 1996.

11. NTSB Aviation Accident/Incident Database, Report No. MIA98LA179, 06/05/1998.

12. Hoddes, E., Zarcone, V., Smythe, H., Phillips, R., Dement, W. C., Quantification of sleepiness: a new approach. Psychophysiology 10:431–436, 1973.

13. Miller, J. C., Fitness-for-duty testing using performance tests in the industrial workplace. Ergonomics in Design 4:2, 11–17, 1996.

14. NTSB Aviation Accident/Incident Database, Report No. SEA00LA046, 02/19/2000.

15. Williams, R. L., Karacan, I., and Hursch, C. J., Electroencephalography (EEG) of Human Sleep, New York, Williams & Wilkins, 1974.

16. NTSB Aviation Accident/Incident Database, Report No. CHI95IA215, 07/09/1995.

17. NTSB Aviation Accident/Incident Database, Report No. SEA98LA002, 10/16/1997.

18. Rosekind, M. R., et al., Managing fatigue in operational settings. 1: Physiological considerations and countermeasures. Behavioral Medicine 21(4):157–165, 1996. Rosekind, M. R., et al., Managing fatigue in operational settings. 2: An integrated approach. Behavioral Medicine 21(4):166–170, 1996. Rosekind, M. R., et al., Fatigue in operational settings: examples from the aviation environment. Human Factors 36(2):327–338, 1994.

19. Ferrara, M., De Gennaro, L., The sleep inertia phenomenon during the sleep-wake transition: theoretical and operational issues. Aviation, Space and Environmental Medicine 71(8):843–848, 2000.

20. NTSB Aviation Accident/Incident Database, Report No. FTW99FA223, 08/16/1999.

21. Moore-Ede, M., The Twenty-Four Society, Reading, Mass., Addison-Wesley, 1993.

22. Wylie, C. D., Shultz, T., Miller, J. C., Mitler, M. M., and Mackie, R. R., Commercial Motor Vehicle Driver Fatigue and Alertness Study: Project Report, Technical Report FHWA-MC-97-002, Washington, D.C., 1996.

23. NTSB Aviation Accident/Incident Database, Report No. MIA97FA232, 08/14/1997.

24. Miller, J. C., Smith, M. L., and McCauley, M. E., Crew Fatigue and Performance on U.S. Coast Guard Cutters, Report CG-D-10-99, Groton, Conn., U.S. Coast Guard Research and Development Center, 1999.

25. Zimmerman, D., Bird, D., Miller, J. D., Doughty, H. E., Sharkey, B. J., Gould, J. E., and Govatski, D., Fatigue and Stress: Fire Season 2000. Report of the Interagency Fatigue and Stress Countermeasures Team. National Aviation Safety Manager, U.S.D.A. Forest Service, National Interagency Fire Center, Boise, Idaho, 2000.

26. AOPA Air Safety Foundation, Joseph T. Nall General Aviation Safety Report, Frederick, Md., Aircraft Owners and Pilots Association, 1999.

27. National Institute on Alcohol Abuse and Alcoholism, Alcohol and sleep, Alcohol Alert No. 41, 1998.

28. http://www.mentalhealth.com

29. NTSB Aviation Accident/Incident Database, Report No. MIA96FA186, 07/15/1996.

Appendix A

Hints for Good Sleep Hygiene

A Healthy Mind in a Healthy Body

- Manage stress as much as possible. Keep things in perspective and focus on what's important. If needed, use relaxation techniques.
- Stay fit. Physical fitness tends to reduce anxiety and insomnia. Even something as simple as brisk walking can have a positive effect, if done regularly.
- Stimulate your mind. Chronic television viewing is associated with poor sleeping. Spend time working, talking, doing chores, and pursuing hobbies.
- Pay attention to healthy nutrition
- Stop smoking. Nicotine stimulates the brain and increases blood pressure and heart rate, disturbing your ability to get to sleep and remain asleep.

Good Sleep Behaviors

- Use a bedtime ritual. For example, read a good book to take your mind off the stresses of work. When you feel drowsy, turn off the light.
- Don't watch the clock. Hide illuminated clocks from view. If needed, set a couple of alarms and arrange for a wake-up call.
- Dress appropriately. Use loose-fitting, soft garments that breathe, in the right weight for the temperature of the bedroom.

A Good Sleep Environment

- Strive for quiet. Low level, consistent sound may be useful. Soft static may help mask unwanted sounds: Set the radio between FM stations and turn the volume to low. You may need to use soft earplugs.
- Strive for darkness. If needed, use a blanket or towel to block a window or the edges of a door. You may need to use an eyeshade.
- Set the room temperature to 65°F (20°C).
- Strive for a humidity level of 60 to 70 percent. At home, you may need a humidifier or dehumidifier. These devices may provide a soft hum of "white noise" that can help mask other noises.
- Strive for security. Add door locks and smoke alarms at home, if needed. Check the door and window locks before sleeping. While traveling, use the door bolts provided by hotels and, if needed, carry a supplemental door security device.

- Design a restful-appearing bedroom at home and keep the bedroom clean and free of clutter.

Good Sleep Equipment

- Use bed sheets that are clean and comfortably soft.
- Use a pillow that allows a healthy sleep posture: on your side with the spine straight or on your back.

Appendix B

Napping Plan

Time at home	GMT (Zulu) time	Elapsed time	Local darkness	Critical tasks	Best nap times

Index

Accidents, 4, 7, 14, 19, 29–30, 36, 43, 44, 68, 101, 127, 152
Air Force Research Laboratory, 23
Aircrew Mission Timeline Tool (AMTT), 70
Alarm clock, 77, 156
Alertness, 3, 6, 98
Ambien® (*see* Zolpidem)
Antidepressants, 138
Antihistamines, 135–136
Anxiety, 14, 24, 120–122, 150, 155
Aserinsky, Dr. Eugene, 63
Attention, 36, 43, 71
  lapse of, 25, 26, 32–33, 37, 49, 51, 66, 112, 147
  sustained (*see* Vigilance)
  withdrawal of, 5, 151
Attitude, can-do, 3, 25
Automation, cockpit, 12, 21, 26, 28–32, 33, 49–50, 112
Autopilot (*see* Automation)

Bali, 101
Birth, maturity at, 63
Body clock, 13, 45, 69, 101, 145, 148
Boredom, 25, 27
Bright light therapy (*see* Fatigue countermeasures, bright light therapy)
Bronchodilators, 137
Brown, I.D., 4, 151

Caffeine, 128–129, 148
Caldwell, Dr. John, 131, 132
Circadian rhythms, 96
  acclimatization, 102, 107, 147
  acrophase, 98, 99
  adaptation (*see* Circadian rhythms, acclimatization)
  central nervous system, 96

Circadian rhythms (*Cont.*):
  effects on performance, 5, 8, 9, 12, 13, 14, 19, 22, 35, 44, 45, 65, 95, 152
  entrainment (*see* Circadian rhythms, acclimatization)
  entrainment cues, 97
  pacemaker (*see also* Suprachiasmatic nucleus), 99
Circasemidian rhythm, effects on performance, 7, 8, 12, 13, 14, 35, 44, 45, 65, 95
Clock, body (*see* Body clock)
Complacency, 32
Control, supervisory, 33
Cortisol, 98
Costs, physiological, 24, 25, 112
Costs, psychological, 24, 25, 112
Crashes, vehicle, 45
Crew duty period (CDP), 5, 6, 8, 9, 19–20, 26, 34–35, 38–39, 111, 112
Crew, scheduling, 3
Cruise control, 29

Decision-making, 20, 21, 26
Dement, Dr. William, 63
Depression, 120–122
Dexedrine® (*see* dextroamphetamine)
Dextroamphetamine, 130–132
Diet (*see* Nutrition)
Diphenhydramine, 135
Diuretics, 137–138
Doxylamine, 135
Drowsiness (*see* Sleep, drowsiness)

Effort, 23–24, 38
  aerobic, 26
  mental, 23–24, 26
  physical, 23–24
Ephedrine, 129–130

Index

Error
  human, 7, 30, 38, 43, 46–47, 51, 52, 83, 95, 147
  meter reading, 44–45
  mistake, 43, 51, 116
  slip, 43, 51, 116

FAA Civil Aeromedical Institute, 23
Factors, multiple, 36
Failure detection, 33
Failure, mechanical, 28
Fatigue:
  acute, 5, 12, 14, 20, 21, 35, 37, 38–39, 67, 111, 112, 143, 149, 152
  chronic, 120–121
  combined, 35
  countermeasures:
    bright light therapy, 14, 104–106, 149
    melatonin, 14, 102–104, 136
    pharmacological, 14, 81, 85–86, 127–140
    planning, 143–152
    prescription, 14, 143–152
  covert, 4, 12, 25, 27, 117–118, 151
  cumulative, 5, 9, 14, 35, 37, 39, 46, 68, 111–120, 143, 149, 152
  defined, 4, 5
  physical, 21, 23, 26
  task-specific, 26–28
  two-peak pattern, 12, 30, 44, 45, 46, 65, 116
  underestimation, 124, 147
Fault diagnosis, 33
Federal Aviation Administration (FAA), 13, 68
Fitness for duty testing, 52

Go pills (*see also* Fatigue countermeasures, pharmacological), 128–132, 139–140
Growth hormone, 62
Guantanamo Bay, Cuba, 19–20
Guilt, 14, 24, 120–122, 150

Halcion® (*see* Triazolam)
Hursh, Dr. Steven, 22
Hydration, 91–92, 148

Incidents, 4, 14, 26, 43, 44, 68
  reporting, 150–151
Insomnia, 77, 155
  sleep maintenance, 61

Jet lag, 7, 9, 12, 13, 20, 95–107, 116, 144, 152
  asymmetric time zones, 106
Judgment, 20, 51

Kleitman, Dr. Nathaniel, 63

Lapse, mental (*see* Attention, lapse)
Little Rock, Arkansas, 35

Maintenance, aircraft, 57–59
Malaise, 4, 26, 152
  due to jet lag, 96, 99
Management, role of, 3, 9, 10
Melatonin (*see* Fatigue countermeasures, melatonin)
Modafinil, 132
Models:
  quantitative, 98
  Fatigue Avoidance Scheduling Tool (FAST), 22
Moore-Ede, Dr. Martin, 101
Moray, Dr. Neville, 33
Motivation, 24, 71

Nall Report, AOPA, 30, 127
Naps, 8, 12, 13, 50, 52, 114, 145, 149–150
  at work, 84–85
  cockpit, 68–70, 114, 145
  length, 70, 71
Narcolepsy, 10
National Aviation and Space Administration (NASA), 13, 68
  Aviation Safety Reporting System (ASRS), 14, 32, 35n, 95
  Ames Research Center, 23
National Interagency Fire Center (NIFC), 116
National Sleep Foundation (NSF), 75, 79, 144, 145
National Transportation Safety Board (NTSB), 19–20, 35, 37, 43, 57, 66, 67–68, 89–90, 113–114, 138
Nicotine, 130, 155

Index

No-go pills (*see also* Fatigue countermeasures, pharmacological), 132–140
Nutrition (*see also* Sleep, nutrition), 15, 86–91
  hunger, 89–90
  plan, 90–91, 146

Obesity, 26

Parasuraman, Dr. Raja, 32
Performance, 25, 27–28, 98
  effectiveness, 22, 36
  mental, 7, 15, 23
  pattern-matching, 34
  physical, 7, 23
Phenergan® (*see* Promethazine)
Phenylpropanolamine, 130
Post-lunch dip (*see* Time of day, midafternoon)
Probability, joint, 36–37, 101
Promethazine, 138
Prozac® (*see* Antidepressants)
Pseudoephedrine, 129–130

Rasmussen, Dr. Jens, 33
Recovery, 7, 8, 36, 62, 98, 112, 143, 152
Restoril® (*see* Temazepam)
Risk, 7, 12, 49
Risk-taking, 26

Safety, 6, 15
Scheduling, 146
SCN clock (*see* Body clock)
Siesta, 14, 115–116, 145, 149–150
Sleep:
  allergy, 78
  alcohol, 86, 133–135, 148
  and flying schedules, 65–68
  bedding, 78, 157
  biology, 12, 59–60
  central nervous system, 60
  cycle, 60–61
  daytime, 46, 60
  debt, 67, 75, 79, 112, 118–120
  deep, 60, 62, 71
  disturbance, 61
  dreaming, 60, 62
  drowsiness, 60
    driving, 81

Sleep (*Cont.*):
  environment, 60, 78, 112, 156–157
  functions, 62
  hunger, 89
  hydration (*see* Hydration)
  indigestion, 78
  inertia, 13, 70–71, 114, 145–146, 148
  latency, 61
  length (*see* Sleep, quantity)
  medication (*see* Fatigue countermeasures, pharmacological)
  nutrition (*see also* Nutrition), 13, 15, 86–91, 148, 155
  nutrition, carbohydrates, 86–87
  nutrition, fats, 88–89, 98
  nutrition, fiber, 87–88
  nutrition, meal size, 89, 98
  nutrition, milk, 87
  nutrition, sugar, 87
  on the job, 25, 26, 27, 82, 112
  pathology, 10–11, 78
  pathology, apnea, 10
  pilots, older, 64
  planning, 148, 155–157
  quality, 12, 112, 144–145
  quantity, 12, 13, 22, 63–64, 65, 75–79, 112, 144–145, 149
  quantity and Internet, 77
  quantity and TV, 77, 155
  quantity, ideal, 82, 118, 119
  recovery, 118–120, 150
  REM, 60, 62–63
  slow-wave, 60
  structure, 60–61
Sleep survey, NSF, 75–86
Sleepiness, 3, 4, 7, 11, 22, 27–28, 66, 78–80, 82–83, 117
  detection, 50, 144–145
  effects on work, 83–84
  injury, 83
Sleepiness Scales:
  Epworth, 79–81, 145
  Stanford, 48, 145
Sleeping quarters, 9
Sonata® (*see* Zaleplon)
Strain, 23–24, 25, 112
Stress, 23–24, 25, 78, 112, 155
  heat, 6
  injury, cumulative, 25–26

Suprachiasmatic nucleus (SCN), (see also Circadian rhythms, pacemaker), 96

Temazepam, 137
Temperature, body, 48–49, 98, 145
Times of day:
  pre-dawn, 7, 8, 12, 30, 36, 37, 43, 45, 46, 47–49, 51-52, 67, 71, 95, 99, 116
  midafternoon, 7, 8, 12, 19, 36, 43, 45, 46, 52, 67, 95, 97, 99, 116
Time off, 8, 122–123
Time zones (see Jet lag)
Triazolam, 136–137
Two-edged sword of human adaptability, 25
Two-peak pattern (see Fatigue, two-peak pattern)

United States Army Aeromedical Research Laboratory (USAARL), 131, 132
United States Coast Guard, 116

Variability:
  human, 28
  lane position, 29
  speed, 29

Vigilance, 6, 21, 25, 28, 30, 31, 37, 50, 66

Wakefulness, 98
  length of , 21–22, 27, 65, 111, 147
  length of and alcohol equivalency, 134–135, 147
Walking, 155
Watch schedule, maritime (see Work, shifts, maritime)
Work demand (see also Workload), 23–26, 111
Work:
  effects on sleep, 84
  irregular schedule, 10, 11, 144
  mental, 6, 50
  night, 9, 10, 11, 50, 67–68, 144
  physical, 6
  safety-sensitive, 6
  schedule equity, 123–124, 146
  schedule predictability, 123–124, 146
  shifts, 7, 68, 79
  shifts, maritime, 116–117
Work-rest cycle, 7, 8, 9, 66
Workday, 7, 67
Workload (see also Work demand), 21

Zaleplon, 137
Zolpidem, 137

About the Author

James C. Miller. Ph.D., is one of the world's leading experts in the science of fatigue and a former fighter pilot in Vietnam.